Forest Lepidoptera Control

THE AUTHORS

Dr. Sathe Tukaram Vithalrao [M.Sc., Ph.D., Sangit Vishard, IBT (Seri.), F.I.S.E.C., F.S.E.Sc., F.S.L.Sc., F.I.C.C.B., F.S.S.I., F.H.A.S.] is presently working as Professor, Department of Zoology, Shivaji University, Kolhapur. He has teaching experience of 31 years in Entomology at University PG department and 18 years in Agrochemicals and Pest Management. He has written 35 books and published 300 research papers in national and international journals of repute. He guided 21 Ph.D. students and completed 6 major research projects (from CSIR, DST, DBT and UGC). He visited Canada (1988), Japan (1988), Thailand (2002, 2004), Spain (2005), France (2005), South Korea (2006) and Nepal (2007) etc. for academic work. He is member of editorial board of eleven prestigious journals. He delivered 35 talks through All India Radio and internal conferences and involved in Doordarshan, S.T.V. and B.T.V. programmes on useful and harmful insects. He published more than 35 popular articles in daily newspapers on insects and sericulture. He got several prestigious awards like "Environmentalists of the Year-2003", "Bharat Jyoti", "Jewel of India", "International Gold Star", "Eminent Citizen of India", "Education Acumen", "Best Educationist", "Eminent Scientist of the Year-2008", "Lifetime Education Achievement", "Lifetime Achievement in Entomology and Insect Taxonomy-2009", Educational Leadership-2011, Asia Pacific International Award-2012, Global Education Leadership Award-2013, etc. He is also working as Research and Recognition (RR) Committee member for Pune University, Pune; North Maharashtra University, Jalgaon; Shivaji University, Kolhapur and DBA Marathwada University, Aurangabad. He has been awarded several fellowships from different scientific and academic societies. He is Chairman of Maharashtra District Environmental Centre of NESA and editorial board member of more than a dozen journals of international and national repute.

Dr. Vishnu Yadavrao Kadam (M.Sc. Ph.D) is Associate Professor in Zoology at Bharati Vidyapeeth, MBSK Kanya Mahavidyalaya, Kadegoan, Dist. Sangali. He has published five research papers and attended and presented several research papers in national conferences/seminars/etc. He is chairman of Satara District Environmental Centre of NESA, New Delhi.

Forest Lepidoptera Control

Dr. T.V. Sathe

Professor
Department of Zoology
Shivaji University
Kolhapur – 416 004, M.S.

Dr. V.Y. Kadam

M.B.S.K. Kanya Mahavidyalaya
Kadegon
Sangli, M.S.

2015

Daya Publishing House®

A Division of

Astral International Pvt. Ltd.

New Delhi – 110 002

Cataloging in Publication Data--DK
Courtesy: D.K. Agencies (P) Ltd. <docinfo@dkagencies.com>

Sathe, T. V., author.
Forest lepidoptera control / Dr. T.V. Sathe, Dr. V.Y. Kadam.
    pages ; cm
    Includes bibliographical references (pages   ) and index.
    ISBN 9789351306320 (International Edition)

        1. Lepidoptera--India--Kolhapur (District)--Identification. 2. Lepidoptera--India--Satara (District)--Identification. 3. Moths--Control--India--Kolhapur (District) 4. Moths--Control--India--Satara (District) 5. Forest insects--Control--India--Kolhapur (District) 6. Forest insects--Control--India--Satara (District)   I. Kadam, V. Y. (Vishnu Yadavrao), author. II. Title.

    DDC 595.78095479   23

Published by   : **Daya Publishing House®**
                             A Division of
                             Astral International Pvt. Ltd.
                             – ISO 9001:2008 Certified Company –
                             4760-61/23, Ansari Road, Darya Ganj
                             New Delhi-110 002
                             Ph. 011-43549197, 23278134
                             E-mail: info@astralint.com
                             Website: www.astralint.com

Laser Typesetting   : **Classic Computer Services**, Delhi - 110 035

Printed at   : **Thomson Press India Limited**

PRINTED IN INDIA

Forest Lepidoptera Control

Dr. T.V. Sathe

Professor
Department of Zoology
Shivaji University
Kolhapur – 416 004, M.S.

Dr. V.Y. Kadam

M.B.S.K. Kanya Mahavidyalaya
Kadegon
Sangli, M.S.

2015

Daya Publishing House®

A Division of

Astral International Pvt. Ltd.

New Delhi – 110 002

Cataloging in Publication Data--DK
Courtesy: D.K. Agencies (P) Ltd. <docinfo@dkagencies.com>

Sathe, T. V., author.
Forest lepidoptera control / Dr. T.V. Sathe, Dr. V.Y. Kadam.
    pages ; cm
    Includes bibliographical references (pages   ) and index.
    ISBN 9789351306320 (International Edition)

        1. Lepidoptera--India--Kolhapur (District)--Identification. 2. Lepidoptera--India--Satara (District)--Identification. 3. Moths--Control--India--Kolhapur (District) 4. Moths--Control--India--Satara (District) 5. Forest insects--Control--India--Kolhapur (District) 6. Forest insects--Control--India--Satara (District)   I. Kadam, V. Y. (Vishnu Yadavrao), author. II. Title.

    DDC 595.78095479   23

Published by   :   **Daya Publishing House®**
                              A Division of
                              Astral International Pvt. Ltd.
                              – ISO 9001:2008 Certified Company –
                              4760-61/23, Ansari Road, Darya Ganj
                              New Delhi-110 002
                              Ph. 011-43549197, 23278134
                              E-mail: info@astralint.com
                              Website: www.astralint.com

Laser Typesetting   :   **Classic Computer Services**, Delhi - 110 035

Printed at   :   **Thomson Press India Limited**

PRINTED IN INDIA

Preface

Western Ghats of Maharashtra, India is among the 18 hot spots of the world for conservation and protection of biodiversity. Pests, diseases and fire are the worst enemies of forest and pests rank first. Among the pest insects, Lepidoptera cause severe damage to forest trees by defoliating, leaf mining and boring stem, roots, fruits and seeds. Diversity, biology and intrinsic rate of increase play an important role in designing suitable control measures hence, above aspects of *Eligma narcissus* Cramer, *Eutectona machaeralis* Walker and *Hypsa product* Moore have been reported in the book. 23 species of Lepidoptera have been described in the book with respect to morphology, distribution and control. Similarly, check list of moths from western Maharashtra have been incorporated. A special chapter for control strategies especially preventive, cultural, mechanical, biological and chemical have also been incorporated in the book. This is unique book on pest Lepidoptera of forest and will be helpful for forest entomologists, environmentalists, students and teachers in India and abroad. Authors are thankful to Shivaji University, Kolhapur and Department of Forestry and Environment, Govt. of Maharashtra for necessary providing facilities to this work and research scientists whose work is cited in the text.

Prof. T.V. Sathe
Dr. V.Y. Kadam

Contents

1

General Introduction

India made satisfactory progress towards economic growth, modernization of the economy and self reliance in various areas and in the most indicators of human development over the past fifty years. Therefore, the average expectancy of life has gone up 60 years in recent years (Dreze and Sen, 1995a).

Due to significant agricultural development India brought under several high yielding varieties of crops, however that require substantial quantities of chemical fertilizers, pesticides and water for appropriate management. India has achieved a considerable degree of food security but, sustainability of Indian agriculture is under threat due to the depletion of groundwater resources. Hence, conservation of natural resources should be given first priority for sustainable agriculture.

The quality of resources like water, soil, forestry etc. have direct relation with the environment quality and life style of human being including other animals. Various ecosystems (aquatic, terrestrial, agro, forest, horticultural etc.) are the parts of environment. The forest ecosystem consists of tree dominated vegetative associations. Forests are good source of timber, fuel, wood, fodder, fiber grasses and non wood forest products which support industrial and commercial activities and also maintain the ecological balance and life support system essential for food production, health and human development. Out of 329 million hectares geographical area, actual forest cover exists only 76.352 million hectares in India. The deforested and/or degraded area in India is much greater. Out of 413 districts, 105 districts had forest cover of more than 33 per cent, 252 districts between 19 to 33 per cent and 226 districts, less than 19 per cent. In 30 districts of India discernible forest cover is lacking.

According to Champion and Seth (1968) Indian forests have following four major groups, *viz.* tropical, subtropical, temperate and alpine. These groups are further divided into 16 subgroups.

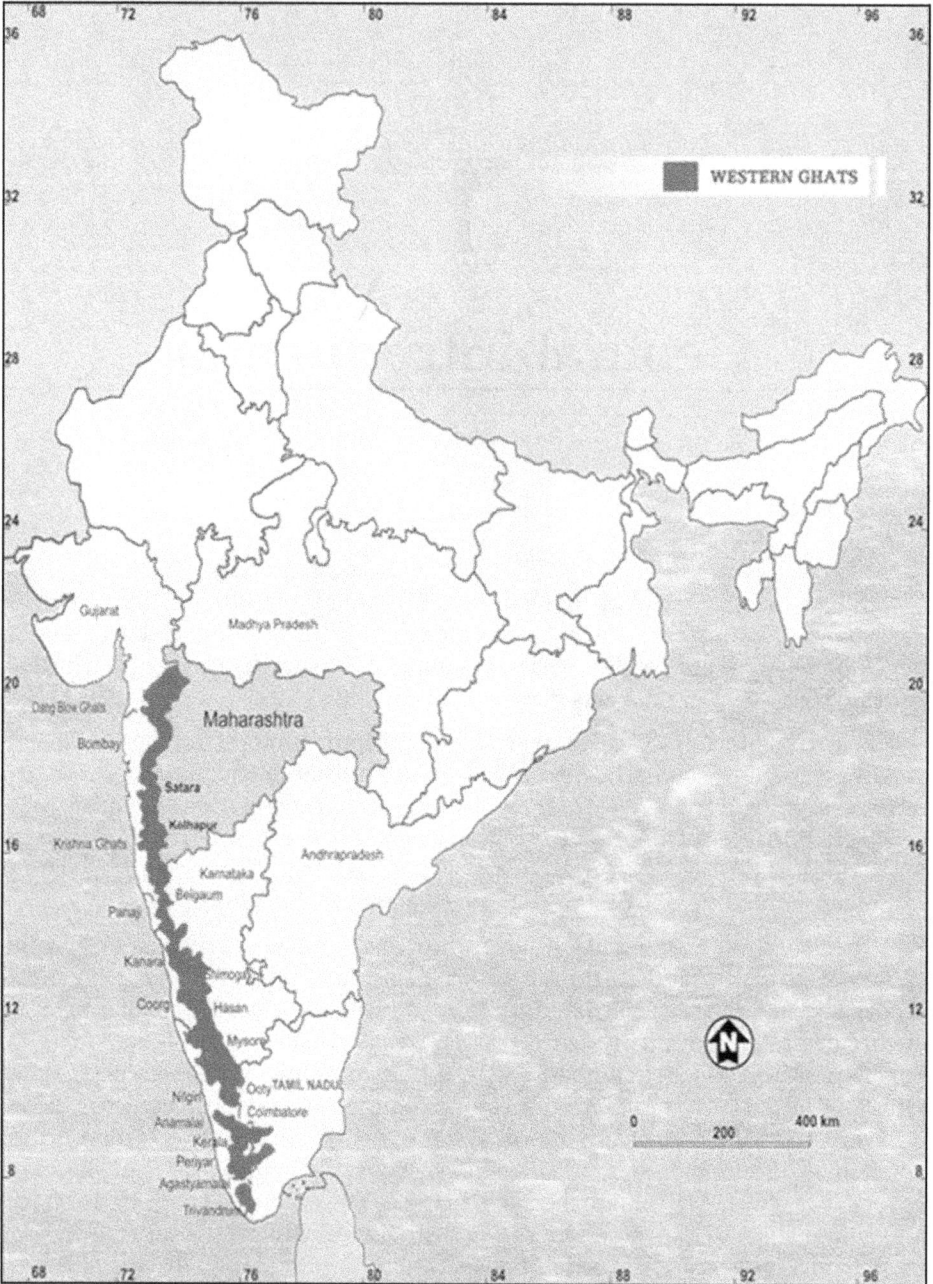

Figure 1: Map of India Showing Western Ghats.

Forest types available in India refer to tropical dry deciduous, tropical moist deciduous, tropical thorn, tropical wet evergreen, subtropical pine, subalpine, Himalayan moist temperate, tropical semi evergreen, montane, littoral and swamp, subtropical broad leaved hill, subtropical dry over green, Himalayan dry temperate and tropical dry evergreen with total per cent of the forest area 38.2, 30.3, 6.7, 5.8, 5.0, 4.3, 3.4, 2.5, 2.0, 0.9, 0.4, 0.2 and 0.1 million hectares respectively.

Indian forest cover losing at the rate of 1,44,000 hectares per year. In 1980 and 1995 it was slowed down to 24,533 hectares. 14 million hectare forest area is lost by India. 0.34 million hectares of forest area was deforested during 1981 to1990, implying a loss of 3.37 million hectares over the period. 64 per cent area of deforestation was reported from dry deciduous forests, tropical rain forests, 11 per cent from moist deciduous forests, 15 per cent from hill and montane areas showed 11 per cent.

The major forests of India are mainly scattered in Himalayan, North eastern and western regions of the country. The western Ghats (Sahyadris) (Figure 2), a prominent mountain range that stretches to more than 1600 km is running parallel to the west cost of India. The western Ghats raise abruptly from sea level to highly dissected plateau up to 2700 metres high and slope to the deccan plains in the east, except at the extreme south. In the Annamalai sector of Kerala the Anamudi peak (2673 metres) is the highest peak in western Ghats. The western Ghats is scattered in states, Tamil Nadu, Kerala, Karnataka, Goa and Maharashtra.

The western Ghats (Figure 2), include high hills west coast of peninsular India from the river Tapi in the north to Kanyakumari in the south. The range is broken at the palghat at the boundry between Kerala and Tamil Nadu. The coastal plain running parallel to the west coast is about 45 km wide till Kanyakumari at the southern tip of the peninsular.However, at some places, it is narrow and the mountains come very close to the sea. Western Ghats is scattered in an area of 1,59,000 sq.km and unique in a humid tropical climate, unusual geological stability and evolutionary continuity of the biodiversity. The western Ghats region of Maharashtra (Figure 3) is characterized by hilly terrain and discontinuous hills and plateau like Bhimashankar and Mahabaleshwar. An unique feature of these hills is the occurance of undulating surface with gentle valley slope at an altitude of 3000 metres, which provide suitable soil and climatic conditions. Therefore, tea and cardamom are cultivated. The coastal plants in the west receives heavy precipitation from western slopes resulting in a series of fast flowing rivers reaching to the Arabian sea. These rivers and their esturies provide extensive fishing facilities and offering potential hydel power. In Maharshtra, the average length of western Ghats is 440 km and it starts from Dhule district and runs southwardly through the districts Nasik, Mumbai, Pune, Satara, Kolhapur, Ratnagiri and Sindhudurga (Figure 3). From Maharashtra Kolhapur and Satara districts of western Ghats (Figure 3) have been selected for the present study.

Kolhapur district has an area of 7,585 sq.km of which forest coverage is above 1,672 sq.km. Out of which 563 sq.km is reserved forest and 417 sq km is protected forest area. In the district the forest area scattered is about 22 per cent. In the Kolhapur district there are 3 types of forests:

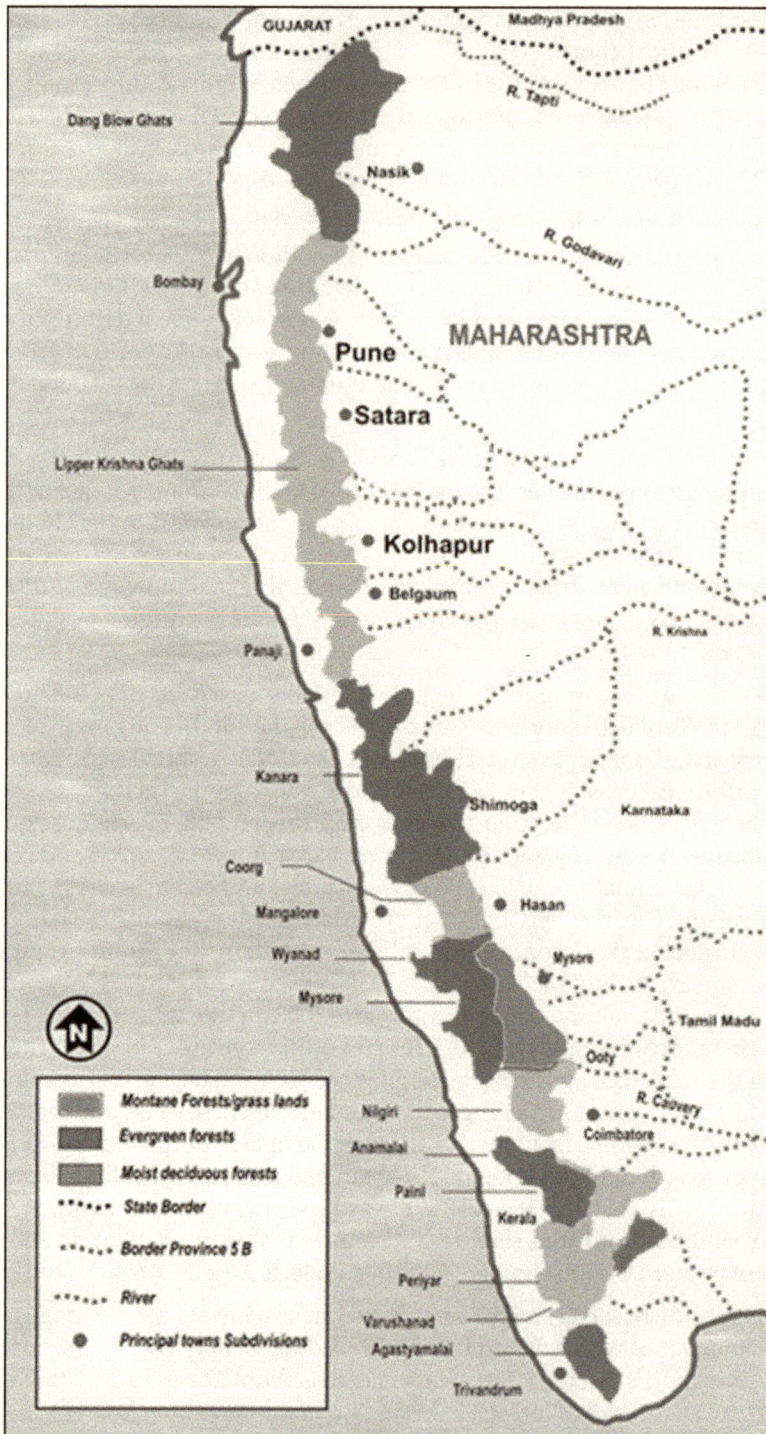

Figure 2: Biographic Subdivision of the Western Ghats.

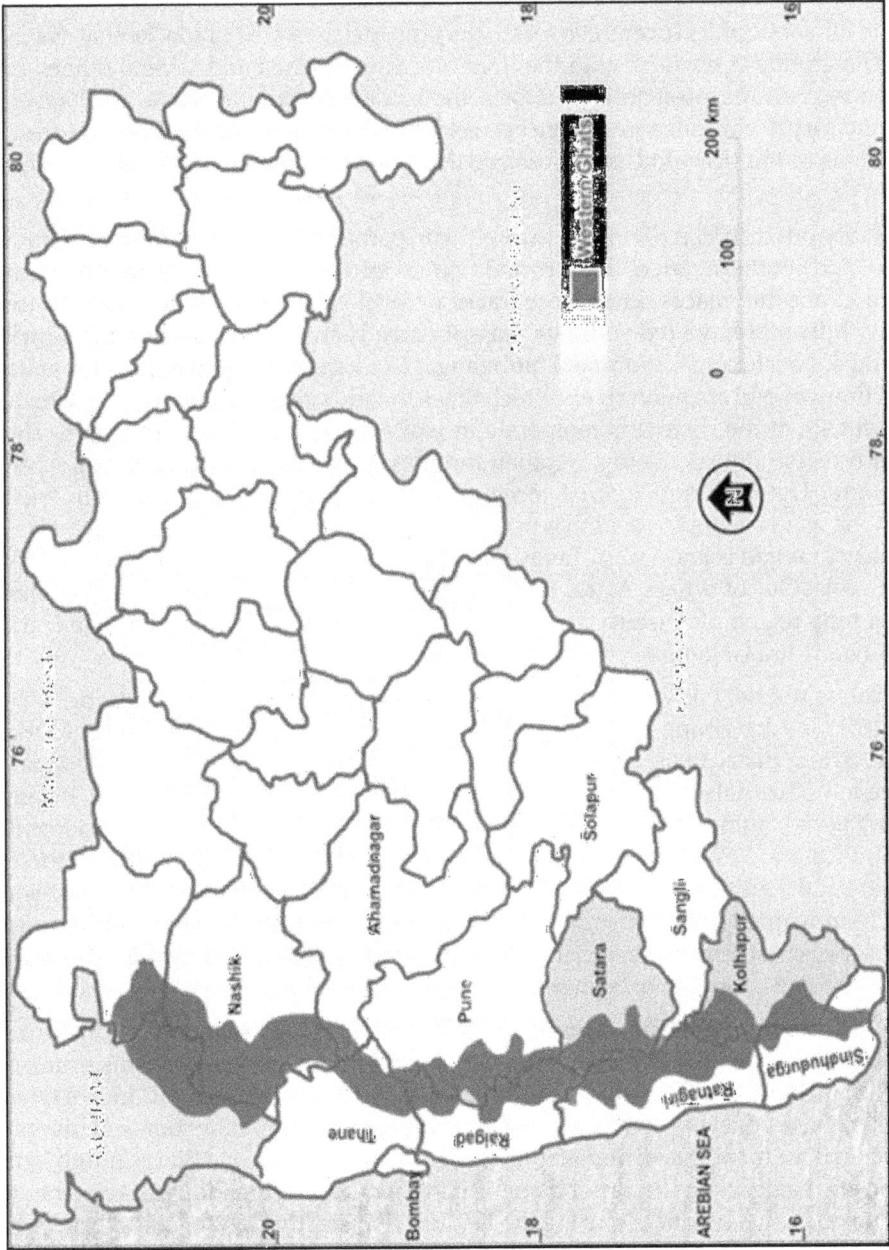

Figure 3: Map of Maharasthra Showing Western Ghats.

1. Subtropical evergreen,
2. The moist deciduous and semi-evergreen, and
3. The dry deciduous forest.

In the subtrophical evergreen forest, the principal trees are Hirada, Jambal, Fanas etc. The ground is covered with the flora of Karvi, Brachen and several others. In semi-evergreen and moist deciduous forest the flora refers Shesum, Bhava, Ain, Umber, Biba etc. Dry deciduous forest have been classified as reserve and protected forest, known as to the firewood and grass are the main marketable products from this forest.

Kolhapur district is divided into three main parts namely Eastern ranges, Central ranges and Southern ranges. Eastern and Central ranges have black soil formed from lava and at some places it has large tracks of fertile land. The western ranges are mostly hilly and have red soil. The majority area in the west is under thick forest coverage. Panchganga, Warana, Dudhganga, Vedganga, Bhogawati, Ghatprabha and Hiranyakeshi are main rivers which flows towards east through western Ghats. The climate of the district is temperate in plains and cool in western Ghats. The Eastern region represents dry weather and it experiences hot winds during April and May. The Kolhapur district receives its major rainfall from the South west mansoon winds. The rainy season is from June to October. The western Ghats receives the heavy rainfall is known as Cherapunji of the Maharashtra. Kolhapur district has 12 tahasils. Out of which Ajara, Radhanagari, Shahuwadi, Gaganbawada comes under hilly region of western Ghats. Therefore, in the present study Radhanagari, Shahuwadi and Gaganbawada have been selected.

Satara district has an area of 10,492 sq.km which lying between latitude 17°5' and 18°1' and longitude 73°33' and 74°74' E. It is bounded by Pune district to the south. Satara district contain eleven tahsils, out of which Maan, Khatav, Phaltan, Koregaon, Khandala and Karad comes under plain region while Wai, Javali, Patan, Mahabaleswhar and Medha comes under hilly region. Satara is also characterized by Ajinkyatara, Yevteshwar, Sajjangad, Potova, Ghatai and Salpani hills. All these hills are at the altitude of 900 to 2150 m from sea level. Khambatki Ghat, Agashiva hills, Mandhardev hill ranges cover dry deciduous forests. Koyana basin hilly range covers west coast semi evergreen forests. In the present study Panchgani, Yevteshwar hills and Koynanagar have been selected.

Biodiversity counts the status of a region or a country. Biodiversity is admixture of plants and animals. Diversity is the variability among living organisms from all sources, including terrestrial and marine. Today's biodiversity is a product of over 3.5 billion years of evolution, involving speciation, migration, extinction and human influence. Due to increased human population and activities, a very large number of species are being lost much faster than the natural rate. A tremendous loss of genetic diversity have been noticed during the last 100 years. Although several species are on the verge of extincition, India is a very good place for bioresources. However, Indian biodiversity is at a very high risk of extinction because of a high degree of endemism. Therefore, it is a global concern for safeguarding biodiversity. India shares 8 per cent of the total number of species found in the world (Khoshoo, 1995).

PLATE 1—Figure 4: Kumbharli Ghat Koynanagar; Figure 5: Koynanagar Forest; Figure 6: Koynanagar Hills; Figure 7: Koynanagar Ghat Matha.

PLATE 2—Figure 8: Panchgani Roadside Forest; Figure 9: Panchgani Hills; Figure 10: Panchgani Thick Forest Area; Figure 11: Yevteshwar Hills.

PLATE 3—Figure 12: Gaganbawada Hills; Figure 13: Radhanagari Dam Side Foerst; Figure 14: Shahuwadi Forest Area; Figure 15: Radhanagari Forest Area.

Indian agro-biodivesity is at seventh place in the world as far as the number of species is concerned and secondary centre of domestication for animal species as horse, goat and several plant species. (Khoshoo, 1995). Their floral wealth is very rich and so is their endemism, including flowering plants. Eastern Himalaya and western Ghats are amongst the 18 hot-spots of the world. Their floral faunal wealth is very rich and so is their endemism.

The western Ghats are home to many endemic, rare and endangered species as well as scores of economically important species and wild relatives of cultivated plants (Chandrasekhara, 1984; Anonymous, 1982; Ananthakrishnan, 1978, 1993). The Southern area shows richest gene pool of the 4500 plant species in western Ghats and 1720 species are endemic. Nearly a one third are rare or threatened and several are extinct, of the 17,000 flowering plant species in India over 4500 occur in the western Ghats (Yadav and Bachulkar, 1995).

Animal biodiversity of western Ghats is also quite rich since a large number of amphibians, fresh water fishes and invertebrates occur in the western Ghats (Joseph, 2004, Daniels, 2000). The western Ghats have very rich flora reported and or described by workers such as Yadav and Sardesai (2002). The timber yielding and wood fuel flora refer to Hopea spp. Nonwood species refer to bamboo, rattan and food value flora *Artocarpus hirutus, Garcinia indica* etc; Edible nuts refer to *Sterculia guttata, Terminalis bellerica* and condiments *Garchinia gummi-gutta, Piper nigrum*, etc. There are several vertebrate animals endemic to western Ghats which are supposed to be important species (Gadekar *et al.,* 1990).

The fauna of western Ghats includes 2100 species of birds (Ali, 1981). *Pantholops hodgsoni, Moschus moschiferus,* etc. are mostly endangered on account of their being hunted for meat and skin. According to Khoshoo (1995), population of some of the above species have dwindled beyond retrieval. The snow leopard *(Panthera uncia)* is the only prominent feline. The brown bear (*Ursus arctos isabellinus*), the black bear (*Selenoroctos thibetanum*), and red panda (*Ailurus fulgens*) are some of the other animals recorded in the western Ghats. The multitired wet evergreen forests of western Ghats have several unique ecosystems which need to study with respect to biodiversity.

About 6700 species of insects have been described from India by 1980 (ZSI, 1983). A large number of insect species have been recorded from Indian forest although little information is available on the characteristics species profile of western Ghats. Some insect groups like the Coleoptera family Passalidae are exclusively found on wet decaying timber on the floor of evergreen forest (Mathew, 1982). Many insects have had their origin from the forest vegetation. In india, the study of forest insects was initiated in 1900 prior to the establishment of Forest Research Institute, Deharadun. F.R.I. Deharadun made significant contribution to the collection and identification of forest insects The above institute is characterized by well established museum of insects and forest materials damaged by the insects.

From India, Beason (1941) summerised 3378 forest insects belonging to the orders Anopleura, Coleoptera, Collembola, Dermaptera, Diptera, Ephemerida, Hymenoptera, Isoptera, Lepidoptera, Neuroptera, Odonata, Orthoptera, Hemiptera. Thysanoptera and Thysanura. According to Nair *et al*. (1981) the estimated number of forest insects is over 16,000 species. Nair *et al*. (1981) studied Lepidopteran bagworms from western Ghats of Kerala. According to Nair (1984); a total of 314 species of butterflies have been recorded from the western Ghats. The insects of silent valley have been studied by the scientists of ZSI in four faunastic explorations between 1979 and 1980.

Joseph (1984) worked on insect life and ecodevelopment of western Ghats. He highlighted the role of insects in the western Ghats ecosystem. Sathe *et al*. (1986-87) studied the fauna of butterflies from western Ghats of Maharashtra. They reported 31 species of butterflies from western Ghats. Similarly, Gaonkar (1996) listed 330 species of butterflies and Bhoje and Sathe (2003) reported 81 species of butterflies from western Ghats.

Satish (1996) recorded 13 species of moths and 16 species of butterflies from Shimoga. Two peirid butterflies, *viz*., *Anaphoeis mesentina* Cramer and *Appias hippo* Cramer have been recordded by Ayyar and Ayyar (1938). *Spindasis elima* Moore was recorded for the first time from Coimbatore. The moths of western Ghats from Maharashtra have been studied by Sathe and Pandharbale (1999). They studied diversity of 139 hawk moths. Sathe and Pandharbale (2001) have contributed on Synthomid, Sphingid and Noctuid moths of western Ghats. The survey of literature indicates that more attention is given on butterflies from western Ghats by various workers (Sathe *et al.,* 1986-87, Goankar 1996, Bhoje and Sathe, 2003 etc.). However, very little attention is paid on the diversity of moths of the region. (Sathe and Pandharbale, 1999; 2004; Pandharbale and Sathe, 2001; 2008 etc.).

Forests are invariably exposed to many biodeteriorating agencies among which insect pests play a significant role. A large population of insects survives on the forests. Most of them are harmful to the growth of plants. They eat leaves, tender shoots, flowers and fruits. They also feed on roots, stems and cause heavy damage in the forests. A large number of insects attack the felled timber resulting in heavy losses. Insects, thus, cause problems to forestry and constraints in the efficient managements of forest resources. The annual losses caused by forest pests to seeds, transplants, standing trees and wood have been computed to be about ten per cent of the total revenue of the forest. (David and Kumarswami, 1982; Negi 1986 and Sensarma and Thakur, 1988).

Teak (*Tectona grandis* Linn.), Shivan (*Gmelina arbores* Linn.), Ain (*Terminalia tomentosa* White and Arn) and Khair (*Acacia catechu* Willd) are some of the major indigenous tree species of Konkan region while Subabul (*Leucaena leucooephala* Boxb.) an exotic species has been recently introduced. The area under these tree species is increasing due to the aforestation programs undertaken by Government and other organisations. Such a large scale planting is likely to be fraught, in the coming years with many insect problems. About 200 species of insects have been reported to attack teak tree (*T. grandis*) (Mathews, 1986) and about 83 species of insects have been reported to attack Shivan (*G. arbores* Linn) in Indian subcontinent (Mathew and Ali, 1987).

Thakur and Pillai (1985) recorded 30 species of insect pests associated with subabul in south India. Among several insect pests attacking forest trees, incidence of teak defoliator (*Hyblaea puera* Cramer) and teak skeletonizer (*Eutectona machaeralis* Walker) has been widespread in the country and its intensity is increased in the Konkan region (Sathe, 2009). Khan *et al.,*(1988) reported 35 to 98 per cent defoliation by *E. machaeralis*, whereas, Ghorpade and Patil (1991) reported 50 per cent damage to teak plantation by the pests in Konkan region. Champion (1934) reported 30 to 65 per cent damage by teak defoliator, *H. puera* Cramer. Sapling borer, *Sahyadrassus malabarious* Moore caused 31.8 to 60.8 per cent damage to teak and eucalyptus (Mathews, 1986 and Nair, 1982). Therefore, there is a need to generate necessary research based data which would help us to protect our forests from insect damages.

Though, the work on the identification of pests of forest trees has been undertaken in other states of the country, no such systematic work has been reported in the State of Maharashtra in general and western Ghats region in particular.

Thus, the work described herein has been taken up with the following objectives.

1. To carry out taxonomy of insect pests infesting forests in western Ghats.
2. To study the biology and intrinsic rates of lepidopterous pests of forest trees in the western Ghats.
3. To control lepidopterous pests of forests.

2

Review of Literature

Linnaeus (1758) Guene'e (1837, 1841, 1853-54), Duponchel (1844-46), Herrich-Schaffer (1845), Packard (1869), Grote (1882, 1890), Tutt (1891-92, 1895, 1896, 1902) and Smith (1891) have published various accounts on the higher classification of Noctuidae. These have followed by Hampson (1892, 1976) who pioneered family level interrelationships. Hampson divided this group into ten subfamilies namely Trifinae, Acontiinae, Gonopterinae, Palindinae, Entelinae, Sarrothripinae, Stictopterinae, Quadrifinae, Focillinae and Deltodinae. In his subsequent publication, Hampson (1902) described several genera under the sub families Trifinae, Quadrifinae, Heliothinae, Headeninae, Cucullinae, Amphipyrinae, Catocalinae, Plusiinae, Pantheinae and Cholephorinae of Noctuidae. He proposed key to the fifteen subfamilies of the family Noctuidae (Hampson, 1902) which was modified by Smith (1904) who recognized seventeen groups namely Habalidae, Eutelidae, Stictopteridae, Hypenidae, Pantheidae, Erebeinae, Catocalinae, Sarrothripinae, Acontiddae, Plusiinae, Erastriinae, Mamestrinae, Xyliniae, Poliinae, Hadeninae, Heliothinae and Agrotinae as belonging to Noctuid moths. Lefroy (1909), Crumb (1956), Forster and Wohlfahrt (1956-1971), Inoue and Sugi (1958-1961) and Todd (1983) have also viewed in order to minimise the confusion regarding the classification of Noctuidae about which Kitching (1984) has reported the present state of confusion, will continue till proper character analysis of different taxa is made. Nye (1975) has suggested fourteen subfamilies for Noctuidae. Workers such as Linnaeus (1758), Fabricious (1793), Guene'e (1852) and Nairender Jit Kaur, 1988) studied, noticed and reported some new species. Swinhoe (1886, 1889, 1891) studied lepidoptera of India and listed 84 species.

Hampson (1892, 1895, 1897) made a significant contribution to the noctuid fauna of India. Hampson (1902, 1912) also published a key to eleven species of genus

Bryophila Tereitschke of the subfamily Acronictinae of British India. Apart from the above works, Hampson (1902-1913) published a series consisting of thirteen volumes, titled "Catalogue of the Lepidoptera phaladae in the collection of the British Museum" which include enough information about noctuid moths fauna of the world. Out of these volumes VII, VIII and IX deals with subfamily Acronictinae (Hampson, 1908, 1909, 1910 and 1913). Sevastopulo (1956), Shull and Nadkerny (1961) and Chatterjee (1967) listed twenty four species from Calcutta and four species from Gujrat under the subfamily catocalinae. Singh (1979) dealt with the moths of different subfamilies of family Noctuidae. He described the genitalia of three species, *mygdon* Cramer, *stolida* Fabricious and *geometricious* Fabricious of the genus *Grammode*.

Sphingidae is an important family of moderate sized to very large moths, including atleast 1,000 species, and widely reported from western Ghats (Bell and Scott 1991; Kitching and Cadiou, 2000; Sathe and Pandharbale, 2004; Sathe and Pandharbale, 2008). Hampson (1892) recorded 121 species from India and Ceylon, and in 1904 the number of known species had risen to 204. Thus, hawk moth fauna of India is therefore very rich. However, recent record indicates that very little attention has been paid on this family. Outside India, Sphingidae has been attemped by Allen (1983), Barlow (1983), Sugi (1987), etc.

Arctiidae as diverse family has been studied from India by Hampson (1976). He visualized four sub familes refer to Arctiinae, Lithosiinae, Nyctolinae and Nolinae. However, very little attension is paid on the family Arctiidae from India except the work of Hampson (1976). Outside India Arctiidae has been attempted by several workers. Noteworthy amongst them refer to Holloway (1988),Watson and Goodger (1986), Edwards (1996), Simmons and Weller (2002), Layne and Kuharsky (2000), Wagner (2005), William, (2009), De Costa and Weller (2005), Dubutolov (2006, 2008), Ferguson and Opler (2007), Duvatolov (2010), etc.

As regards to the family Hypsidae,taxonomical picture of India is not fair. This family has been attempted by only Hampson (1976) from India. He reported 4 genera and 28 species under this family from India.Hypsidae is close with Arctiidae but an areole on fore wing can separate this family from Arctiidae.

Review of literature indicates that work on survey of insect pests of forest trees was carried out by many scientists. Beeson (1913) reported that *H. puera* and several species of Arctiids such as *Aularches miliaris* L., *Teratodes monticollis* Grey and *Spilosoma obliqua* Walker defoliate teak forests.

Garthwaite (1939) reported that *Calopepla leayana* Latr. was a serious defoliator of *Gmelina arborea* Roxb. in Assam, Bengal, Mumbai and Chennai. Similarly, the leaves of *T. grandis* were seriously damaged by *Hapalia machaeralis* Wlk. and *H. puera* and estimated losses were upto 13 per cent. In Madhya Pradesh, *Celosterna scabrator* Fab. was a most notorius pests of babul, *Acacia arabica* and teak plantations (Beeson, 1939, 1941).

In India and Myanmar serious outbreaks of *E. machaeralis* were recorded towards the end of the growth season of teak between September and January. Infestation was also noticed between April and June in southern and central India.

According to Mathur (1941) several species of noctuiids have defoliated forest plantation, among which *Paeotes subapiocaslis* Wlk., *Beara nubiferella* Wlk., *Maurilia iconica* Wlk., *Dinumma placens* Wlk., *Pleurona falcate* Wlk., *Spodoptera litura* Fabr., *Phytometra chalcytes* Esp., *Tiracola plagiata* Wlk., were observed on teak; *Garella rotundipennis* Wlk., *Ophiusa tirrhaca* Cram., *S. litura, Pseudelyna rufoflava* Wlk., *Selpa seltis* Moore, *Risoba obtructa* Moore, were found infesting *T. tomentosa* and *Heliothis armigera* Hubn., *Pandesma quenavadi* Guen., *Pericyma umbrina* Cram. on *A. catechu, Selpa celtis* Moore, a serious pest of *G. arborea* and *T. tomentosa* was commonly noticed after rains in Assam, Dehra Dhun and Ratnagiri, North Thane and Palghar region of Maharashtra.

Mathur and Singh (1960, 1961) reported various pests of *A. catechu* Willd. Which could be grouped as defoliaters *viz. Diapromorpha turcia* Fabr., *Myllocerus catechu* Marshall, *Trigonophorus hookeri* White, *Traminda mundissima* Wlk., *Eriocela inagulata* Guenee, *Pericyma umbrina* Guenee, *Eriboes athmas athmas* Drury; as sap feeders *viz., Laccifer lacoa* Kerr, *Lecanium longulum* Douglas and as wood borers *viz. Lyctoderma abigum* Leshe, *Sinoxylyon* sp.Lesne, *Demonax bueteae* Gardner *Xystrocera globosa* oliver and *Anthaxia acacia* Thery. They also reported several insects on *Terminalia* sp. Among which *Euproctis bipunctanex* Hamp. and *Apoderus tranquebaricus* Fabr. were the most important. They further reported various pests on teak among which defoliators, *H. puera* Cram. and skeletonizer, *E. machaeralis* Wlk. caused severe damage to the teak plantations. Netawiria and Tarumingkeng (1971) reported *E. machaeralis* as the serious pests of teak. They also reported that *E. machaeralis* was noticed since 1968 in the main teak growing area of Java.

In Kerala, Nair (1982) reported the teak sapling borer, *Sahyadrassus malabarious* a serious pest of teak, eucalyptus, Gmelina and Albizzla plantations. The borer has a wide host range and attacked more than forty species of shrubs and trees belonging to twenty two families in the forest ecosystems.

Vaishampayan and Bahadur (1983) studied the seasonal activity of teak defoliator *H. puera* and skeletonizer *E. machaeralis* under light trap and reported that presence of *E. machaeralis* moth was observed from July to December with a peak from August to October. Patil and Thontadarya (1983) also studied the seasonal abundance of *E. machaeralis* larvae throughout the year. Pearce and Hanapi (1984) reported, *Acherontia lachesis* as a serious pest of teak in Malaysia. Larvae of the pest were found rapidly defoliating teak. Thakur and Pillai (1985) surveyed the subabul nurseries and plantations in Andhra Pradesh and Tamil Nadu and recorded 30 species of insects associated with subabul including order lepidoptera. All lepidopterous pests were defoliators.

Bhowmick and Vaishampayan (1986) studied migratory behaviour of *H. puera* in India. Singh (1986) reported that bagworm, *Cryptothelia crameri* caused serious defoliation to *A. catechu* and *A. nilotica* plantations in Haryana, Punjab and Himachal Pradesh. Reddy *et al.,*(1988) reported the tussock caterpillar, *Dasychira mendosa* a polyphagous pest causing economic damage to *Terminalia arjuna* and *T. tomantosa.*

Khan *et al.* (1988) studied the seasonal activity and abundance of *H. purea, Euproctis* sp. and *E. machaeralis* in forest. Sensharma and Thakur (1988) listed various insect pests of teak, including lepidoptera such as *S. malabaricus* Moore, *H. puera,* and *E. machaeralis.* Beeson (1941) studied the biology of teak skeletonizer, *E. machaeralis.* David and Kumarswami (1982) reported that female of *E. machaeralis* laid 250 to 500 eggs on leaves and as many as 10 to 12 generations were completed in a year.

While studying the biology of teak skeletonizer, *E. machaeralis.* Patil and Thontadarya (1987) reported that there were five larval instars in the pest life cycle with a distinct prepupal period. Joseph (1984, 2004) studied the insect life development in western Ghats. Joseph *et al.,*(1983) studied the outbreak of hairy caterpillar *Eupterote* sp. on cardamom from south India.

Kehimkar (1997) contributed on Indian moths of forest and plain region. Similarly, Mathew (1990) studied lepidopterous fauna of forest from India.

Pandharbale and Sathe (2001) contributed on new species of *Syntomis* from western Ghats of Satara region. Sathe and Pandharbale (1999) studied biodiversity of sphingid moths from western Maharashtra including Ghats. They reported 55 species of sphingids from the region. Sathe *et al.,*(1997) studied wild silkmoth diversity and their wild host plants from western Ghats. Sathe and pandharbale (2001) also contributed on a new speices of the the genus *Syntomis* from Ghats. Sathe and Pandharbale (2004) studied biodiversity of moths from western Ghats of Satara with respect to host plants, seasonal abundance and features of moths. They reported 127 species of lepidoptera from the western Ghats of Satara.

Recently, Sathe and Pandharbale (2008) studied 26 lepidopteran pests of forest with respect to pest features, damage caused by the pests to forest trees, host plants and control strategies. Very recently, Sathe (2009) contributed on the biology, damage, host plants, distribution and control stratigies of lepidopterous pests of 21 forest trees, *viz.*, Teak, Deodar, Shisham, Ain, Bomboo, Sal, Babul, Neem, etc. from westerm Ghats. He studied biology and control of *H. puera, E.macheralis, S.malbaricus, Ectropis deodarae, Geometrina* sp, *Plecoptera reflecta* Guenee, *H. armigera, Notolophus antique, Trabla vishnu, Lymantria obfuscate, Lymantria mathura,* etc. He also emphasized biological control strategies for above said pests as ecofriendly control. Review of literature indicates that lepidopterous pests are not very widely attempted in previous days and there is need to concentrate on research on various disciplines of lepidopterous insect pests.

Intrinsic Rates of Increase

The estimates of the rate of growth of the pests have tremandous importance in pest management (Howe, 1953). In a given environment an individual living animal shows its own characteristics qualitatively and quantitatively at longevity and fecundity. The value of development, are determined in part by the environment and in part by inherent characteristics of the living animal itself. According to Thompson (1924) the inherent characteristics of the animals are collectively called the "innate capacity for increase". He visualised the first mathematical method for population dynamics. Later, Lotka (1925) derived the function for "the intrinsic rate of natural increase" and then Birch (1948) used the same for animal ecology and for the insect populations. In the present study the life tables were constructed according to Birch (1948) as elaborated by Howe (1952) and Watson (1964).

Review of literature indicates that life table studies have been attempted in different orders of insects by several workers, noteworthy amongst them refers to Morris and Miller (1954) on *Choristoneura fumiferana* (Lepidoptera), Stark (1959) on *Recurvaia starki* (Lepidoptera), Richards and Waloff (1961) on *Phytodecta olivacea* (Coleoptera), Le Roux *et al.*,(1963) on *Spilonota ocellana* (Lepidoptera); Waloff (1968) on *Sitona recansteinansis* Herbst (Coleoptera) and on *Arytacina cenistae* (Homoptera), Mcleod (1972) on *Neodiprion swainei* Midd. (Hymenoptera), Tamaki *et al.*,(1972) on Zebra caterpillar (Lepidoptera), Bains and Shukla (1976) on *Chilo partellus* (Swinh.) (Lepidoptera), Bilapate and Pawar (1980) and Reddy and Bhattacharya (1988) on *Helicoverpa armigera.*

3

Lepidoptera Preservation
and Collection

Accurate methodology plays very important role in avoiding the errors in scientific results. Any minor change in material or methodology, leads in drastic change in the results. Therefore, for maintaining accuracy following materials and methods were adopted in completion of the work.

1. Glass Cages (Figure 16)

Two cages of quadrangular shape with size 25×25×30 cm were adapted for rearing of insects. Each cage consisted with wooden case and glass walls on three sides. One side door was made by muslin cloth with a sleeve for handling the insects and for rearing of caterpillars of lepidopterous forest pests.

2. Glass Troughs (Figure 17)

The glass troughs of size 9 × 20, 12 × 25 cm were used for rearing caterpillars collected from forests ecosystems. In the glass troughs, leaves of host plants were placed with caterpillars and covered with muslin cloth for protection and prevention of escape of caterpillers during rearing.

3. Plastic Containers (Figure 18)

Plastic containers of size 18 × 6cm, 13 × 5 cm, 12.5 × 4.5 cm and 4 × 4 cm in diameter and height respectively were used for handling and rearing the caterpillars Containers were perforated with small holes for aeration to insects.

4. Petridishes (Figure 19)

Petridishes of size 18.5 cm and 9 cm in diameter were used for rearing of caterpillars and pupae.

PLATE 4—Figure 16: Glass Cage; Figure 17: Glass Trough; Figure 18: Plastic Containers; Figure 19: Petridishes; Figure 20: Spereading Board; Figure 21: Test Tubes.

5. Spreading Board (Figure 20)

Spreading boards were used for pinning the moths.

6. Test Tubes (Figure 21)

The test tubes of size (maximum) 15 × 2.5 cm have been used for handling small moths.

7. Rearing Jars (Figure 22)

For storing the moths temporarily in the laboratory and for mating, oviposition jars of size 17 × 6 cm and 12.5 × 1 cm (height and diameter) were used. Each glass jar was covered with muslin cloth for good ventilation to insects and for avoiding escape of insects.

8. Insect Net (Figure 23)

The insect net of following description was used for collection of moths from the field. The insect net with aluminium handle nearly 15 inch in length having a circular metal ring of 6 inch in diameter and a collecting bag of 30 inch in depth made up of ordinary mosquito netting cloth was attached to the iron ring.

9. Insect Boxes (Figure 24)

The standard insect boxes made by Vaman D. Purohit and Sons of varying size (44 × 29 × 7.5 cm, 45 × 30 × 7 cm, 29.5 × 22.5 × 4 cm) have been used for preservation of moths. The insect box was made up of wooden portion covered with soft wood for pinning the insect and the upper side was with glass and rest of the sides were made up of wooden material.

10. Insect Cabinet (Figure 25)

Insect cabinet of size 7.5 × 4.5 × 3.4 feet, made by Vaman D. Purohit and Sons was used for keeping the pinned moths. Each cabinet contain 24 boxes. For avoiding fungal and micro insect infections, naphthalene bolls were kept in each box at one corner.

11. Oven (Figure 26)

For drying insects oven was used.

12. Camera (Figure 27)

Photography of insects was made in the field/laboratory conditions with the help of canon camera (made in Japan) having macro/micro lenses (1:5 ×, single lens reflex). For close up photographs of moths a close up lens of 55 mm +1, +2, +3, +4 and 55 mm macro was used. The moths have been photographed when they found sitting on various plant parts.

1. Collection and Preservation of Moths

From the western Ghats of Kolhapur and Satara district moths have been collected with the help of insect net. The collected moths were anesthetized for short period for their description. After making description the species were released in the

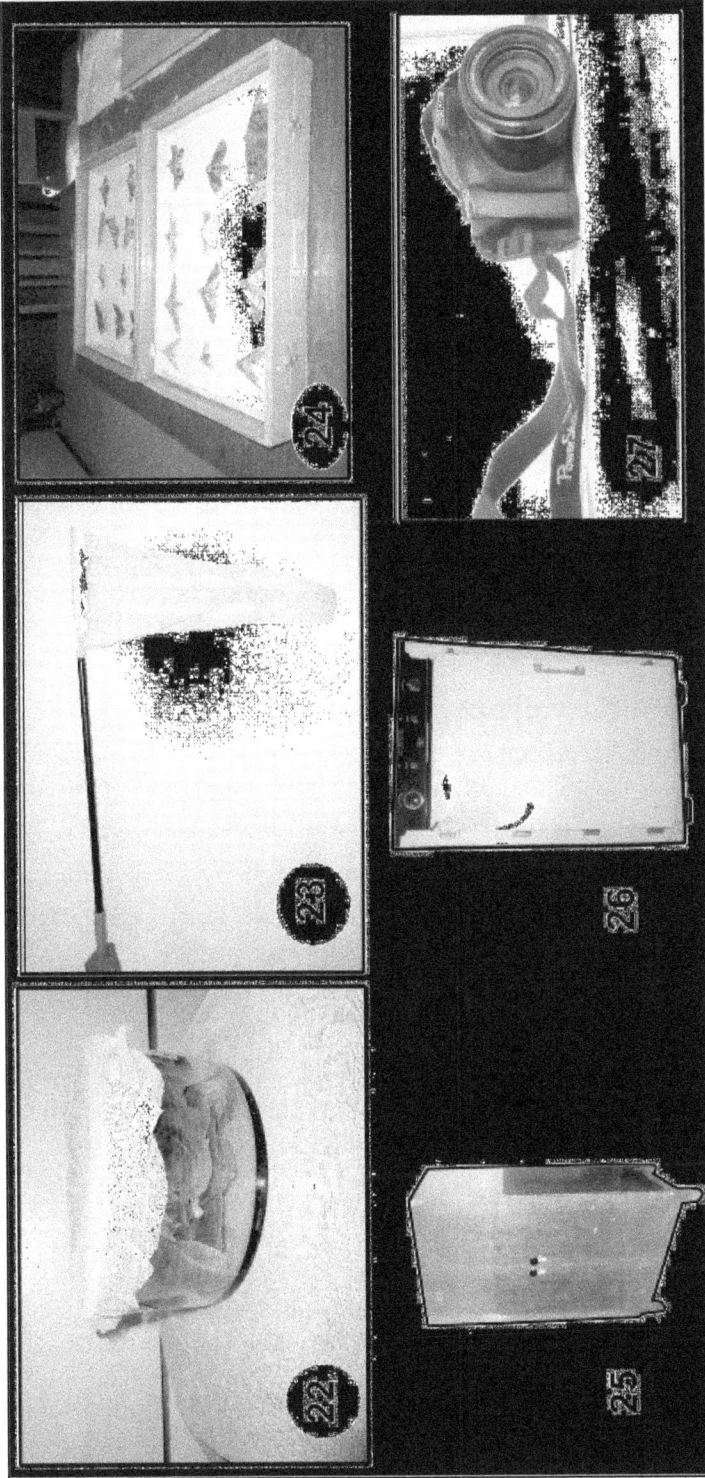

PLATE 5—Figure 22: Rearing Jar; Figure 23: Insect Net; Figure 24: Insect Boxes; Figure 25: Insect Cabinet; Figure 26: Oven; Figure 27: Camera.

environment from which they were collected. Only few and abundant species were preserved in the box by pinning appropriately. The pinned insects were dried in the oven at 60° and only intact and scientifically pinned insects were preserved in insect boxes.

2. Collection and Preservation of Larvae/Caterpillars

Caterpillars have been collected from various plants from western Ghats of Kolhapur and Satara district and reared in the laboratory on the natural food material for emergence of moths. After emergence of moths,species have been studied biosystematcally. Either laboratory reared or field collected moths have been considered for taxonomical descriptions.

3. Taxonomy of Moths

Taxonomical studies have been made on moths with the help of compound microscope/binocular/nacked eye. Measurements of species and body parts were taken in mm in the description. With the help of camera lucida sketches of some body parts have been drawn. Coloured photographs of moths have also been taken as supportive description of the species (Figures 39–130). The taxonomical description of the moths have been made as per the terminology of Hampson (1976) and Bell and Scott (1991). The species described are time being preserved in the department of zoology, Shivaji University, Kolhapur and will be deposited in ZSI, Kolkatta.

4. Survey and Survelliance

Survey of moths have been made at 15 days interval from selected study spots of western Ghats of Kolhapur and Satara districts. Following spots were fixed for study.

Kolhapur district—Radhanagari,Shahuwadi and Gaganbawada.

Satara District—Panchgani, hills, Yevateshwar hills, and Koyananagar.

The species have been noted at various spots in study area through out the course of study. Observations were made on the species by one man one hour search method at each spot (100 sq. ft) at evening.

5. Host Records

Host records of lepidopteran pests have been prepared by observing larvae feeding on specific plants in the western Ghats of study area. Later, the flora and feeding fauna have been identified by consulting appropriate literature.

6. Distribution of Moths

Distributional record of larvae/moths have been made from study spot by observations of the species, at 15 days interval.

7. Seasonal Abundance

Seasonal abundance of moths have been studied by spot observations of the species is from different study spots at 15 days intervals through out the year during course of study by adopting one man one hour search method.

8. Biology of Lepidopterous Pests

During present studies, following, 3 species were abundant and found causing serious damage to the forest trees. Hence, they were selected for biological study.

1. *Eligma narcissus* Cramer.
2. *Eutectona machaeralis* Walker.
3. *Hypsa producta* Moore.

A brief account of the technique employed in these studies is given below.

The initial culture of the pests was obtained by collecting larval stages from the various forest/plantations during the course of study and reared in the laboratory in 20 cm height and 10 cm diameter glass jars and other equipments. Leaves of respective host plants were provided as a food for the larvae and it was changed periodically till the pupation. Freshly emerged moths from the culture were sexed and released for mating in mating chamber. A twig of host plant having 2 to 3 healthy leaves were kept in conical flask containing water for oviposition. The mouth of conical flask was plugged with cotton. The flask was kept in glass jar. The top of the glass jar was covered with muslin cloth secured firmly with a rubber band and the outer side of the jar was covered with black carbon paper for providing darkness. Cotton swab soaked in five per cent sugar solution was kept suspended in the glass jar as a food for moth. Food material was changed daily. Eggs laid by females were removed with moist camel hair brush and kept in 10 cm diameter petridishes containing the pieces of moist blotting paper at the bottom. After hatching, the larvae were reared in the glass jars on respective host food plants leaves, Thus, mass culture was maintained in the laboratory and various aspects of biology were studied by using this culture.

Mating

A newly emerged female and two male adults were released in glass cage of size 25×25×25cm (Figure 16). The outer side of which were covered with black carbon paper and mouth of the cage with muslin cloth. The cotton swab soaked in five per cent sugar solution was kept suspended in the glass cage as adult food, which was changed daily. A total of 6 such glass cages were kept under observations.

Preoviposition Period

The period required from the emergence of female moth from pupa to the commencement of egg laying was recorded for ten females. The average preoviposition period was worked out.

Oviposition Period

The period during which the female moth laid eggs was recorded for ten females and the average oviposition period was worked out.

Postoviposition Period

The period from ceasation of oviposition by a female moth till its death was recorded for ten females. The average postovipositon period was worked out.

Fecundity

One female and two male adults were released in glass cage Figure 16 as described above. Such glass cages were taken into account for observations. A twig of host plant kept in conical flask filled with water and the mouth plugged with cotton swab. The flask was kept in a glass cage for oviposition, Observations on egg laying were recorded daily. The total number of eggs laid by a female were counted every day upto the death of female. The mean number of eggs laid by a female moth was worked out on the basis of total number of eggs laid by ten females and the data were presented.

Morphometrics of Eggs

The morphometrics of eggs were recorded by using ocular micrometer scale. The average was worked out on the basis of data of twenty eggs.

Incubation Period and Hatching Percentage

Eggs laid overnight were transferred into 9 cm diameter petridish containing moist blotting paper at its bottom. Ten eggs were maintained per petridish and twenty such petridishes were kept under observations. Eggs were observed daily for hatching and observations were continued for a fortnight. The average incubation period and hatching percentage were worked out from the data and presented.

Larval Rearing

Ten newly hatched larvae were transferred into 9 cm diameter petridish with tender leaves of host plants. The larvae were transferred with the help of moist camel hair brush.

A piece of blotting paper was provided at its bottom. Fresh healthy leaves were provided as food for the larvae until pupation. The left over food and excreta were removed daily. Twenty such petridishes were maintained for study.

Larval Instars

Larvae were observed to study the instars. The exuviae seen either in the petridish or in the excreta indicated the change in instar. Larval instar was also noted by changes in head width. The durations of larval instars were recorded and described. The linear measurements on head width, body length and body breadth were recorded using micrometer scale and millimeter scale. The data thus, obtained were averaged and presented.

Prepupal Period

The period required from initiation of construction of cocoon by larva to complete formation of pupa was recorded for twenty larvae. The average prepupal period was worked out from the observation and the data were presented. The average length and breadth of prepupae were studied.

Pupal Period

Twenty prepupae were kept under observation in glass jar till the adult emergence and the period required for pupal stage was recorded. The average pupal period was worked out. The mean length and breadth of pupae were also taken by using millimeter scale.

Adult Longevity

Both sexes of the newly emerged adult from the culture were separated and kept in 20 cm height and 10 cm diameter cylindrical glass jar, the top of which was covered with muslin cloth secured firmly with a rubber band. The cotton swab soaked in five per cent sugar solution was kept suspended in the glass jar with the help of thread which served as a food for adults. The longevity of male and female adults was recorded by counting the duration between the emergence and the death of the adult. Twenty adults were considered for the average longevity of male and female adult and the data were presented.

Adult Morphometrics

The measurements on body width, body length and wing expanse were recorded by using millimeter scale for twenty adults of both sexes separately. The body width was recorded across the thorax while wing expanse was recorded by spreading the wing completely and measured horizontally. The data thus obtained were averaged and presented.

Sex Ratio

Ten pupae were kept in each glass jar and immediately after they were formed, ten such glass jars were kept for the studies. The adult emerged were separated for their sexes and the sex ratio was worked out on the basis of the number of male and female adult emerged from the total number of pupae.

9. Intrinsic Rates of Increase

Birch (1948)visualized the following equation to study intrinsic rate of natural increase.

$$\sum e^{-r} m^{x} l_{x} m_{x} = 1$$

where,

'e' is the base of the natural logarithms,

'x' the age of the individual in days,

lx the number of individual alive at age, 'x' as a portion of one, and m_x the number of female offsprings produced per female in the age interval 'x'.

The sum of the products lx mx is the net reproductive rate,

'R_0' which is the rate of multiplication of the population in each generation measured in terms of females produced per generation.

The approximate value of cohort generation time 'Tc' was calculated as follows:

$$T_c = \frac{l_x m_x X}{l_x m_x}$$

The formula : $\quad r_c = \dfrac{\log_e R_o}{T_c}$

provides the arbitrary value of innate capacity for increase 'r_c'.

This was an arbitrary value for rm and value of r_m upto two decimal places was substituted in the formula until the two values of the equation were found which lies immediately above or below 1096.6. The two values of

$$\sum_x e^{7-r} m^x l_x m_x = 1$$

were then plotted on the horizontal axis against their respective arbitrary r_m s on the vertical axis. The point of intersection gives the value of r_m accurate to 3 decimal places. The precise generation time 'T' was calculated as

$$T = \dfrac{\log_e R_0}{r_m}$$

and the finite rate of increase (λ) was calculated as-

$$\lambda = e^r m.$$

Adult moths of *E. narcissus*, *E. machaeralis* and *H. producta* were reared under laboratory conditions (25 ± 2°C, 65 ± 5 per cent R.H., 12 hrs. photoperiod). The laboratory culture was used for determining intrinsic rate of increase. Rearing details are given in the biology of moths.

The life tables were prepared with the help of fecundity data and later the intrinsic rates of natural increase of population of moths were calculated. All the experiments were carried out at laboratory conditions (25 ± 2°C, 65 ± 5 per cent R.H., 12 h photoperiod) and replicated for ten times.

10. Checklist of Moths

Checklist of moths from western Ghats of Kolhpaur and Satara district have been made by giving visits to selected spots in the study area. The moths have been observed for every 15 days and records have been made on. their availability, characters and host plants. Many moth species have been time being collected with the help of insect net for observation of characteristics of the species and they have been released immediately in the environment after observations of the characteristics. Some of the moth species have been observed resting on the certain host plants were also considered for preparing checklist. Observation on larvae of certain species and some time rearing of them to an adult emergence was also the part of preparing checklist. Observations for checklist have been made at evening and night at 15 days interval.

11. Toxicity of Pesticides

It was studied under laboratory condition by treating caterpillars of various species with pesticides.

4

Diversity of Lepidoptera

Insects are the largest group of arthropods and also in the entire animal kingdom. Insects are divided into 29 orders (Imms, 1957). The class insecta is subdivided into two subclasses, Apterygota and Pterygota. Pterygota is further subdivided into Hemimetabola and Holometabola. The butterflies and moths are grouped under the order Lepidoptera of subdivision Holometabola. From the view of evolution, moths and butterflies were among the last to arrived on the evolutionary scene, around 160 million years ago, following the evolution of flowering plants. According to Kehimkar (1997) there are about 25,000 known species of butterflies and over 1,20,000 moths. Some primitive moths exist even today and are believed to be ancestere of caddisflies. The early moths had bitting mouth parts. Strangely, some moths such as the Atlas and Tasar moths have no mouth as they do not feed at all at the adult stage. Moths differ from butterflies by absence of clubbed antennae, franulum in wings, mostly they fly by night and are comparatively dull coloured than butterflies. They also keep their wing parallel to the sitting substrate.

The mouth parts of moth are long coiled tongue as proboscis. The Lepidoptera shows four distinct stages of their life *viz.*, egg, larva, pupa and adult. The order Lepidoptera is characterized by

1. Medium to large sized,
2. Covered with overlapping flat scales forming colour pattern,
3. Mouth parts modified into coiled sucking proboscis,
4. Two pairs of wings,
5. No cerci.

Mani (1993) divided Lepidoptera order into three suborders namely,

1. Jugate,
2. Rhapalocera and
3. Frenatae.

☆ **Jugate** : In this case, wings interlocked by jugum, veination of fore and hind wings similar.

☆ **Rhapalocera** : In this case antennae are knobed at tip or thickned before tip, without frenulum but with strongly arched and butterflies are included under this suborder

☆ **Frenatae** : In this case antennae are simple or variable rarely swollen at the tip and with frenulum.

The suborder Jugate is divided into following superfamiles, Micropterygoidea and Hepialoidea.

The sub order Frenatae provides following superfamilies namely,

Cossocidea	Castnioidea	Zygaenoidea
Incurvarioidea	Nepticuloidea	Pyralidoidea
Sphingoidea	Elachistoidea	Gelechoidea
Yponomeutoidea	Tortricoidea	Ptgerophoroidea
Tineoidea	Uraniodea	Geometroidea
Drepanoidea	Noctuoidea	Saturniodea
Bombycoidea		

and Rhoplacera shows two superfamilies namely Hesperoidea and Papilionoidea.

In the present work, families namely, Arctiidae, Sphingidae, Hypsidae and Noctuidae, were selected for their taxonomical diversity, distribution and their host plant records.

The family Noctuidae is recognized by areola, Sc free at the base and shortly fusing with R. M2 often reduced or absent. Under the family Noctuidae, about 25,000 species are recorded from different parts of world (Kitching, 1984; Pemm, 1993).

Morphological Consideration

Characteristics of a typical moth have been represented in Figure 28.

Head (Figures 29 and 30)

The dorsal skeleton is divided by two transverse sutures into clypeus (2) epicranium and occiput. The epicranium forms laterally the sockets for the antennae. The clypeus is the largest plate of the three. It bear at the anterior margin of the labrum. The labrum is cariniform tubercle fronted over the base of the tongue concealing the medial part of the epistome.

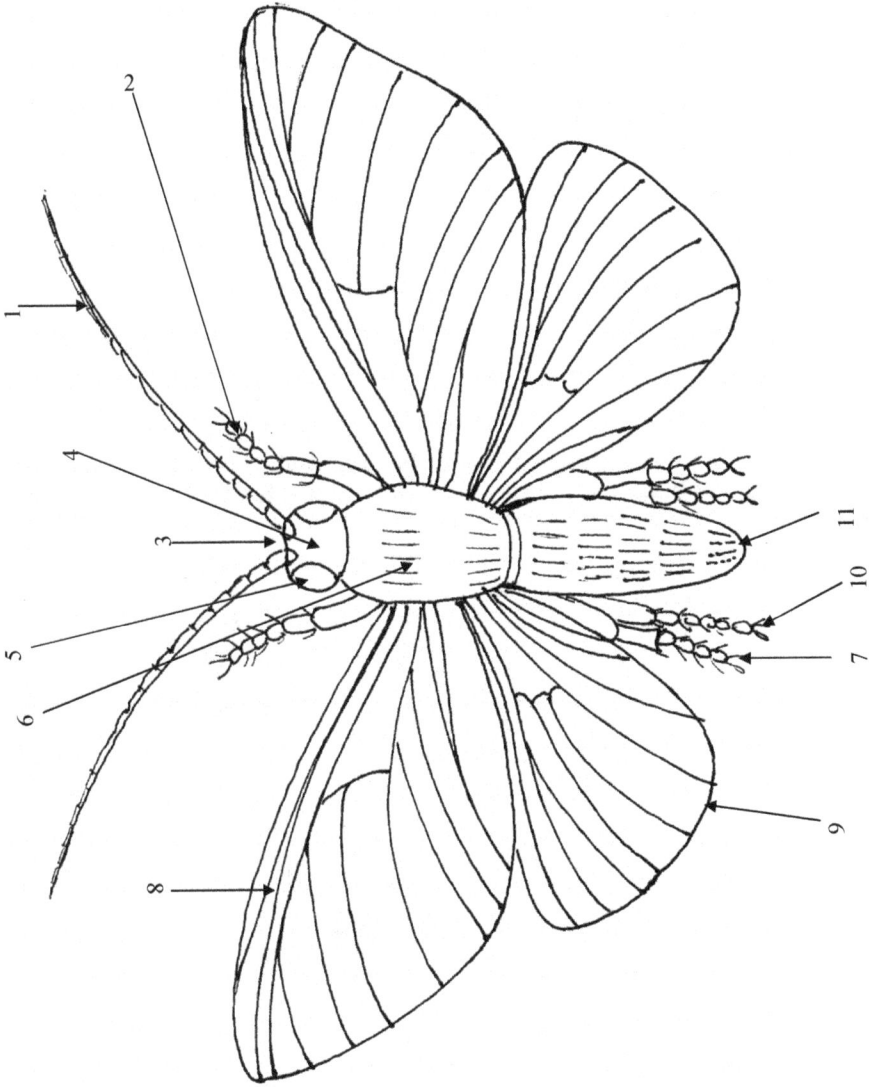

PLATE 6—Figure 28: Adult moth 1. Antenna 2. Fore leg 3. Mouth 4. Head 5. Eye 6. Thorax 7. Middle leg 8. Fore wing 9. Hind wing 10. Hind leg 11. Abdomen.

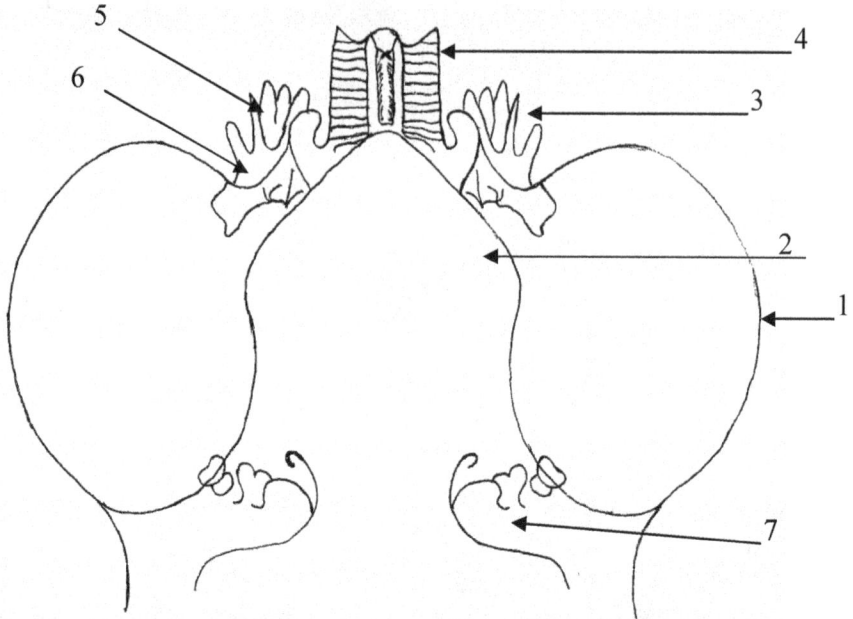

PLATE 7—Figure 29: 1. Morphology of Head 1. Head of adult moth - denuded, dorsal view. 1. Eye 2. Clypeus 3. Maxillary palpus 4. Tongue 5. Pilifer 6. Genal process 7. Base of antenna.

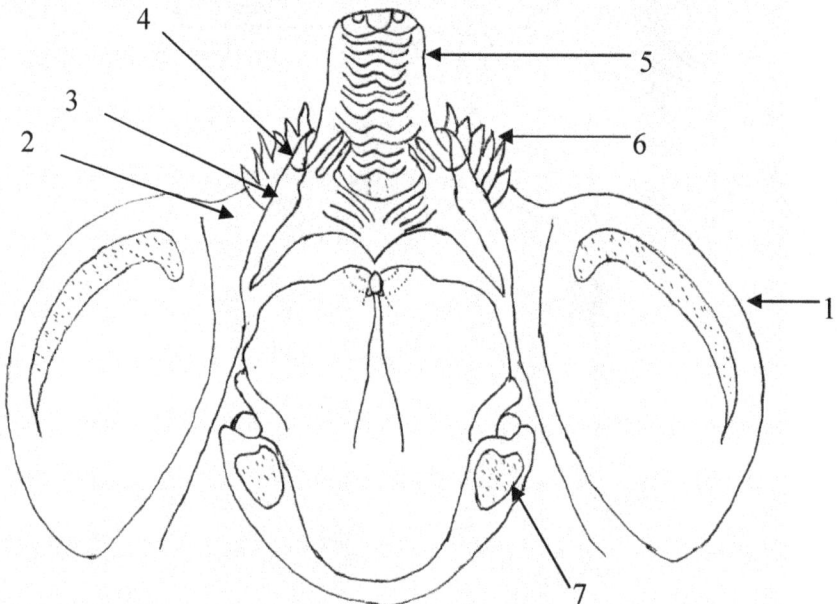

PLATE 7—Figure 30: Head of adult moth – denuded- ventral view, labial palpi removed. 1. Eye 2. Genal process 3. Maxillary palpus 4. Pilifer 5. Tongue 6. Maxillary palpus 7. Groove in which labial palpus is inserted.

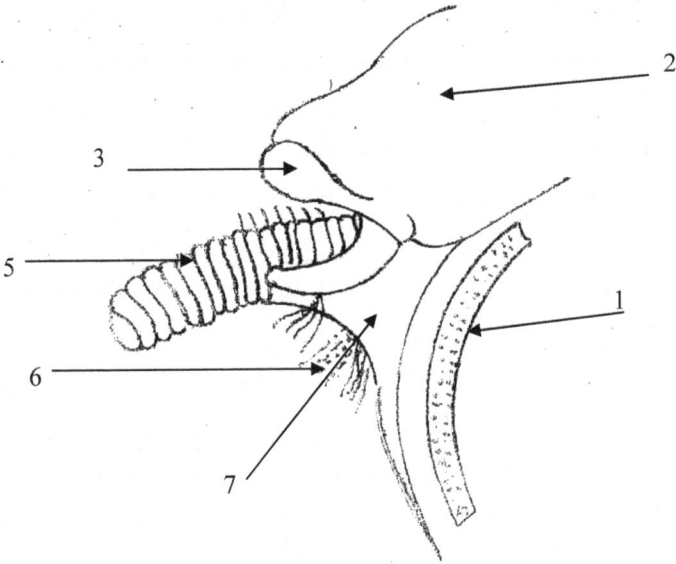

PLATE 8—Figure 31: Mouth Parts of Adult Moth: Lateral view -1. Eye 2. Clypus 3. Labrum 4. Proboscis 5. Maxillary palpus 6. Pilifer 7. Genal process.

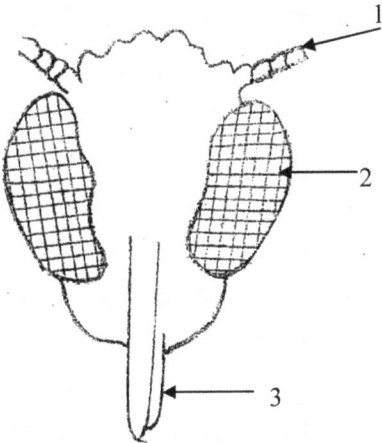

PLATE 8—Figure 32: Head of adult moth – Frontal aspect 1. Antenna 2. Eye 3. Tongue.

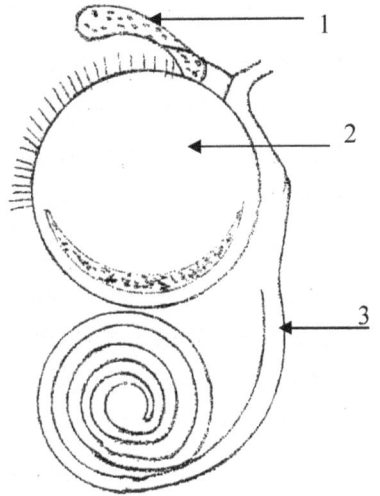

PLATE 8—Figure 33: Head of adult moth lateral aspect 1. Antenna 2. Eye 3. Proboscis.

The epistome covers the base of tongue. The lateral processes are designated as "pilifer". The palpus is large broad in lateral aspect, closely contiguous to the head and has a short mini segment. Other parts are shown in Figures 31–33. Eyes are subglobular and vary in size, never hairy, covered above by a kind of eye brow and below by a large tuft of hair.

Thorax (Figures 34 and 35)

Mesonotum, composed of the prescutum, scutum and postscutum. The parascutum triangular in dorsal view. The sentum (1) is widest behind and a little longer than broad. The postscutum varies in size and shape. Metanotum contain similar parts. The scutum (3) is divided into two halves. The postscutum (3) is narrow. The sternum and peristernum (12) are not completely separated from one other. The peristernum is large at plate extending obliqualy dorsal and mesiad from the meral stuture (13) separating the meral and sternal parts of the sternite to the membrane connecting meso and prothorax. Below the parasternum there is episternum (11), hyposternum (15) is fused with episternum and the marginal strips along the coxal cavity. The episternum is obliqualy truncate with the upper inner angle more or less pointed. The meral half of the sternite is made up of the paramerum (8) and the protomerum (13) two more or less strongly convex plates, together with the large eipimerum (10). A marginal strip (14) situated along the meral cavity is separate by a more or less distinct suture from the epimerum. The metasternite is more simplified than the mesosternite.

Leg (Figures 36 and 37)

The leg comprises the coxa, trochanter, femur, tibia and tarsus. The coxa is inserted in a groove formed by sternal parts of the sternite. The tronchanter (troch) is borned by the coxa and supported by coxa and is supported behind by merum. The femora always remain simple. Tibia and tarsus undergo several modifications. The apex of the fore tibia bears a strong process (a thorn). Tibiae are more or less spinose. The mid tibia has one pair of slender spurs, but the proximal pair very often disappears. The hind tarsus is generally longer than the mid tarsus. The claw segment is composed of the claw (onychium), the false claw (paranychium), the pad (pulvillus) and the empodium. The empodium is a small tubercle above the pad between the claws bearing one bristle.

Abdomen (Figure 34)

Abdomen is ten segmented. The ninth and tenth of male and the eighth to tenth of female are modified. The sphingid abdomen possess an armature of spines. Abdominal segments consists of a tergite (at) and a more or less triangular lateral plate, the parapleurum (pp), it bears no trace of real spines. The first abdominal stigma (sti) lies free in the membrane in front of the para-pleurum. The second to sixth tergites are longer with the sides more strongly covering anal in most species. The eight tergite is small and partly (male) or completely (female) concealed by the seventh. The parapleura of segments 2 and 8 are membraneous and bear the stigma. The second stigma is situated upon the tergite and third one half upon the tergite and half

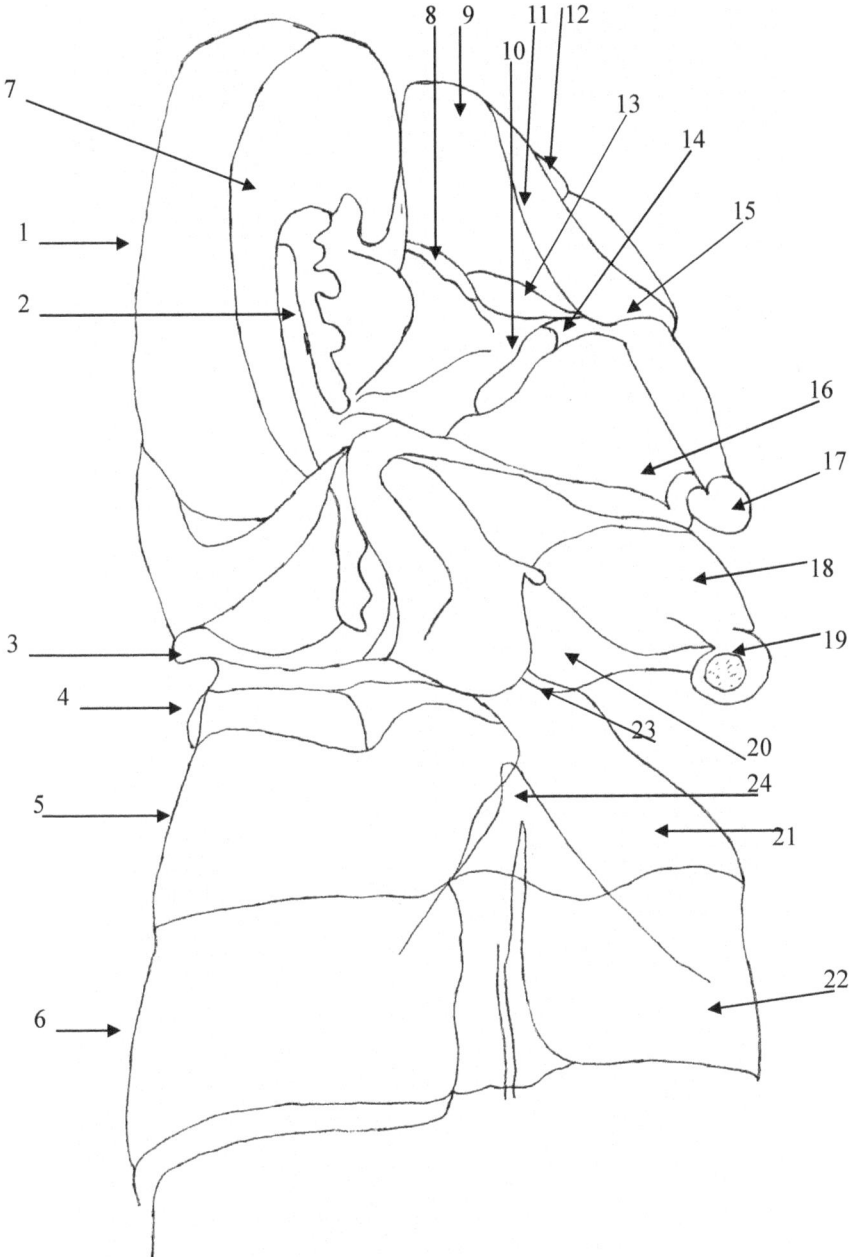

PLATE 9—Figure 34: Morphology of thorax with abdominal segments. 1. Scutum of mesothorax 2. Insertion of wing 3. Scutellum of mesothorax 4,5,6 Targites of first, second and third abdominal segments 7. Mesothoracictegula 8. Paramerum 9. Parasternum 10. Marginal strip 11. Episternum 12. Peristernum 13. Paramerum 14. Marginal strip 1. Episternum 13. Parapleuum 14. Marginal strip 15. Hyposternum 16. Merum 17. Trochanter 18. Coxa 19. Troch 20. Merum 21, 22 Sternites of second and third abdominal segments 23. Hypomerum.

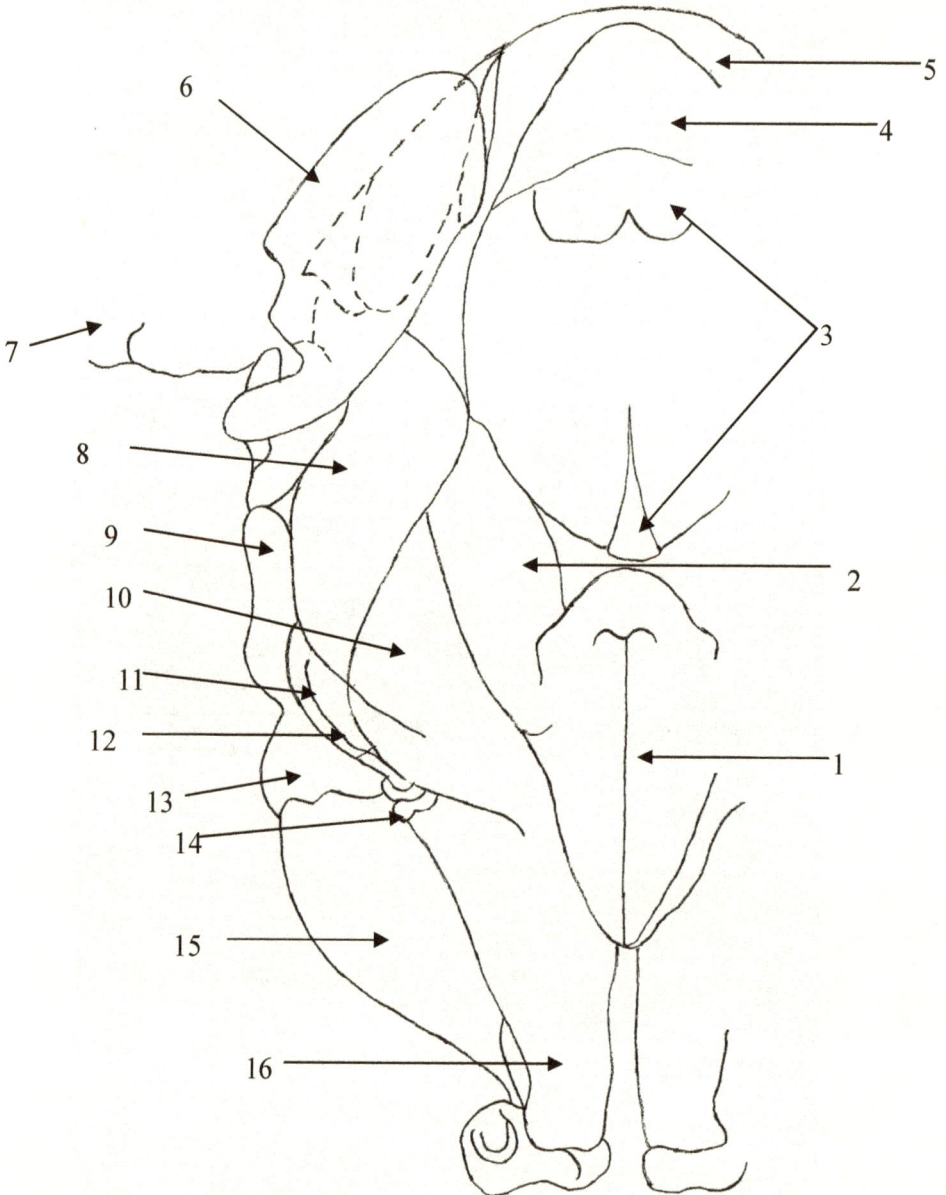

PLATE 10—Figure 35: (B) Morphology of thorax 1. Sternum 2. Peristernum dose 4. Prescutum 5. Scutum 6. Mesothoracic tegula 7. Anterior wing : endose, skeleton 8. Parasternum 9. Paramerum 10. Episternum 11. Parapleurum 12. Hyposternum 13. Epimerum 14. Trochantinus 15. Merum 16. Coxa.

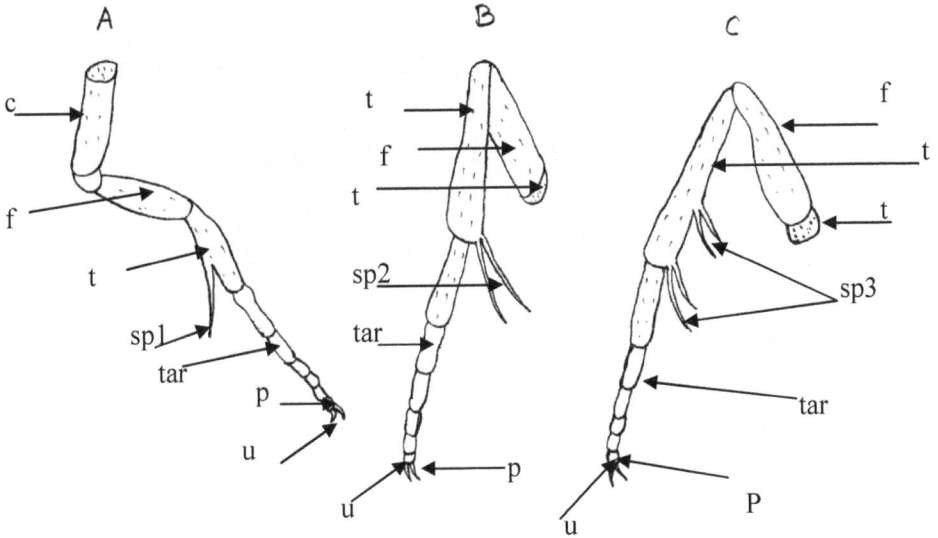

PLATE 11—Figure 36 : Legs of moth A. Froe leg c. Coxa, t – Trochanter f- Femur, t-Tibia, tar – Tarsus B- Middle leg u – Ungues, p-Pulvillus, t- Tibia, Sp. 1 Single anterior spur, sp. 2 Paired medial spurs, sp. 3 Two paris of posterior spurs C. Hind leg f- Femur.

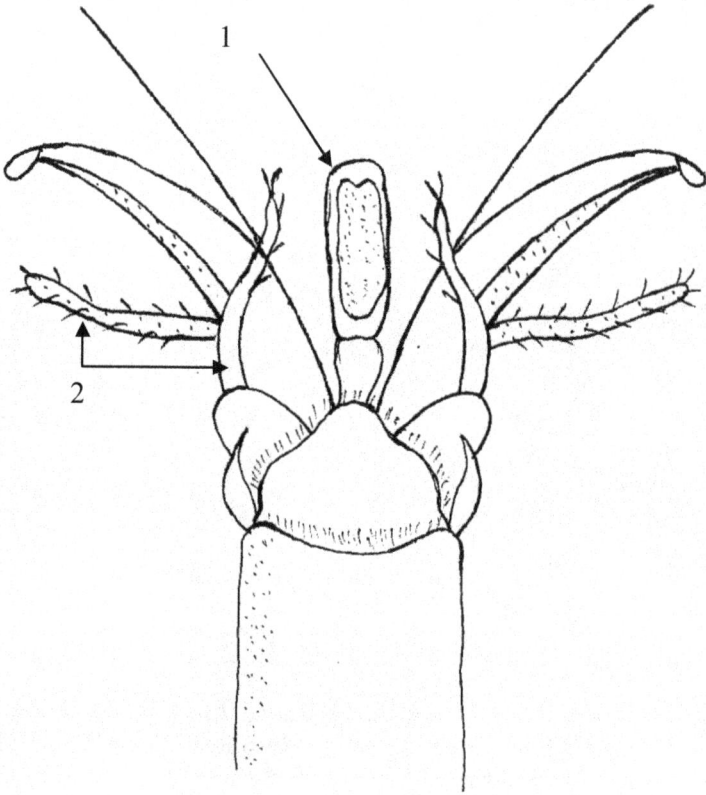

PLATE 12—Figure 37 : Leg of Moth Tarsus Ventral aspect 1. Pulvillus 2. Paronychium.

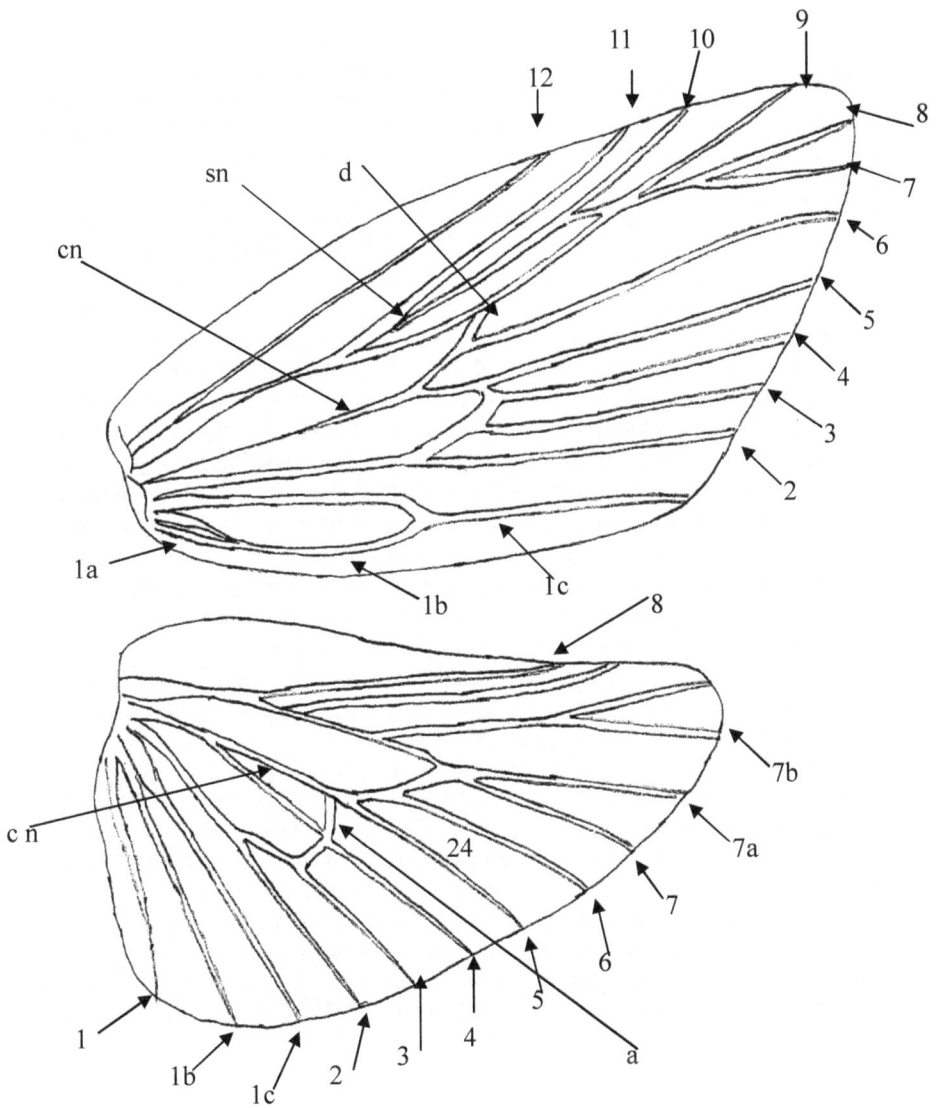

PLATE 13—Figure 38: Fore wing and Hind wing of moth (Diagramatic).

c.n. : veinlet in cell, the fork of which forms the discocellulars d.s.n. : Stalk of veins 10; 10 anastomosing with 7 and 8 to form the areole. II. Hind wing; 7,a, b, c, d, five subcoastal nervules.

upon the parapleurum. The sternites of the first and last segments undergo remarkable modifications.

Wings (Figure 38)

The details of wings are shown in Figure 38. The frenulum and retinaculum are sometime reduced vestigial or absent. The fore and hind wings are very variable in shape in moths.

Genitalia

The copulatory apparatus of the male is composed with the ninth and tenth segments. The ninth segment or tegument is a strongly chitinized girdle, broadest above and minute basally. The belt is ventro-laterally dialated into a large flap (cl), the clasper or value and bears the harpe (H). The pleurum is attached to a proximal strip of chitin (pl) and to the sternite. The tenth segment (x.t. and x.u.) stands in very close connection with the ninth. Between the sternite and tergite is the anus (A) and between the tenth sternite and the ninth the penis funnel (P-F.), from which protrudes the penis sheath/aedeagus (p). The tenth tergite bears stiff hairs. The clasper is normally sole shaped with dorsal and ventral margins rounded. There are various modifications by reductions and by division and the development of a special armature. In order to examine the female organs it is necessary to remove the seventh to tenth segments. The vaginal armature, lying hidden in a cavity in most species, must be pushed outwards by pressure from the inside to become plainly visible.

FAMILY : ARCTIIDAE

Arctiidae is a large and diverse family which contain 11,000 species world wide. This family is recognized by a tympanal organ on metathorax and a pair of glands near ovipositor and anal glands in females.

Moths are mostly nocturnal or crepuscular habits and with narrow elongate fore wings or in some genera with short broad wings, Whilst Nyctemera and its allies are diurenal. Palpi usually short and porrect frenulum present. Fore wings with vein la separate from 1b; lc absent. Hind wings with vein la and 1b; 1c absent, vein 8 arising from 7 genarally at or beyond the middle of the cell. Larvae with all the prolegs clothed with hairs and forming a cocoon for pupation.

KEY TO THE SUBFAMILIES

a. Thorax and abdomen stoughtly built   1. Arctiinae

b. Thorax and abdomen slender.

     a′ Fore wings without tufts or lines of raised scales.

     a″ Fore wing not produced at apex   2. Lithosiinae

     b″ Fore wings slightly Poduced at apex   3. Nycteolinae

     b′ Fore wings with tufts or lines of raised scales   4. Nolinae

SUBFAMILY : LITHOSINAE

Moths of diurnal or crepuscular habits; in the typical section the fore wing is very long and narrow and both wings have lost some of their veins. Proboscis present. Mid tibia with one pair of spurs, hind tibia generally with two pairs. Larva sparsely covered with hair and forming a slight cocoon. Most of the typical section being lichen feeders.

Key to the Genera

1. Fore wing long and ample, the outer margin obliqua.

   Fore wing with an areole,

   Fore wing with the areola long,

   Palpi upturned, the second joint reaching vertex of head.

   Palpi with 3rd joint long and spatulate   *Eligma*

2. Palpi with the 3rd joint much shorter and acute. Fore wing
   with apex and outer margin rounded   *Migoplastis*

3. Fore wing with the apex produced the outer margin straight   *Dilemera*

4. Palpi with 2nd joint not reaching vertex of head. Hind wing with
   vein 3 from some way before the end of cell   *Curoba*

   Hind wing with vein 3 from the end of cell   *Argina*

GENUS *ARGINA*

Argina, Hubn.verz. P 167 (1818)

Type, *A. cribraria,* Clerck.

Range : Africa, Mauritius, China and throughtout India, Shri Lanka, Myanmar, Andamans, New Guinea and Australia.

The genus *Argina* Hubner is visualised by following characters. Palpi upturned, reaching the vertex of head. 3rd joint short. Antennae cilialed in both sexes, Mid and hind tibia with minute terminal pairs of spurs. Hind wing of male with a fold on inner margin containing a glandular path near the base, with a tuft of long hairs beyond it, the anal angle produced to a point. Fore wing with vein 3,4,5 from close to angle of cell, 6 from upper angle, 7 and 10 from a long areole formed by the anastomosis of 8 and 9. Hind wing with veins 3,4,5 from the angle of cell, 6 and 7 from upper angle, 8 from middle cell.

From India 3 species of *Argina* have been reported.

KEY TO THE SPECIES OF GENUS *ARGINA*

1. Head, thorax and fore wing brownish red/scarlet   3

   Head, thorax and fore wing pinkish brown.

2. Head, thorax and fore wing orange yellow or white   5, 6

3. Head, thorax and fore wing with scarlet or brownish red. Collar with
   two black spots ring yellow   *Argina argus*
   collar with two black spots ringed grey. Hind leg tibia
   smaller than femur   *Argina indica*

4. Costal line with 4 black spots   *Argina lidhayssus*

5. Costal line with 5 black spots   *A. indica*

   Hind wing shallow tail like with black spot   *A. cribraria*

   Hind wing with swallow tial like without black spot   *A. kolhapurensis*

6. Colour of Head, thorax and hind wing almost pure white   *A. dulcis*
   Colour of Head, thorax and hind wing yellow   *A. paradalina* and *A. satarensis*

   Basal region of wing forms circle of 6 black spots, Hindwing with
   4 large black spots   *A. satarensis*

   Basal region of wing not forming circle of black spots   *A. paradalina*

7. Colour of head, thorax and wings bright yellow. 5 black spots
   on marginal lines of hind wing   *A. guttata*

Argina kolhapurensis sp.nov. (Figure 39)

Female

Body 20 mm long, 4.9 mm broad, Fore wing 28 mm long, Hind wings 20 mm long, wing expanse 60.9 mm, antenna 6 mm long, Hind leg 20 mm long.

Head

Orange yellow, 2mm long, 2.1 mm broad; eyes large, black round; frons black, genae yellowish. Labrum minute and attached to clypeus; palpi short and porrect; Frenulum present. **Antenna :** Ciliated, 6 mm long, scape black, 0.4 mm long, pedicel black, 0.3 mm long, flagellum blackish 5.3 mm long, many segmented.

Thorax

4 mm long, 3 mm broad, orange yellowish with central black spot. Collar yellow. **Fore wing :** 28 mm long, 13 mm broad, pale pinkish, 26 white oval spots with central black patch present. **Hind wings** : 20 mm long, 15 mm broad, bright yellow, 7 black spots present, vein 1a and 1b present, 1c absent. **Hind leg** : Yellow, 13.2 mm long, coxa 1 mm long, trochanter 0.2 mm long, femur 5 mm long, tibia 4 mm long, tibial spurs present. Tarsus 3 mm long.

Abdomen

14 mm long, 3 mm broad, bright yellowish with dark black spots on lateral and mid dorsal side.

Male

Similar to female except sexual characters and smaller in size.

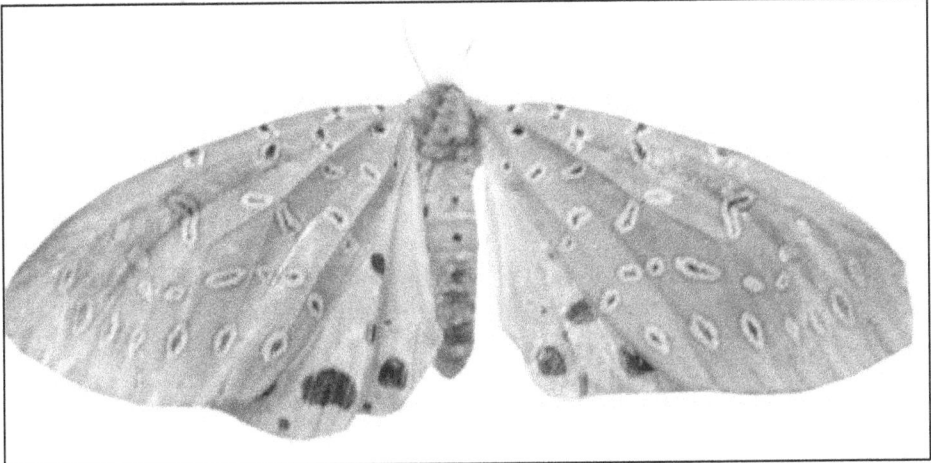

PLATE 14—Figure 39: *Argina kolhapurensis* **sp.nov.**

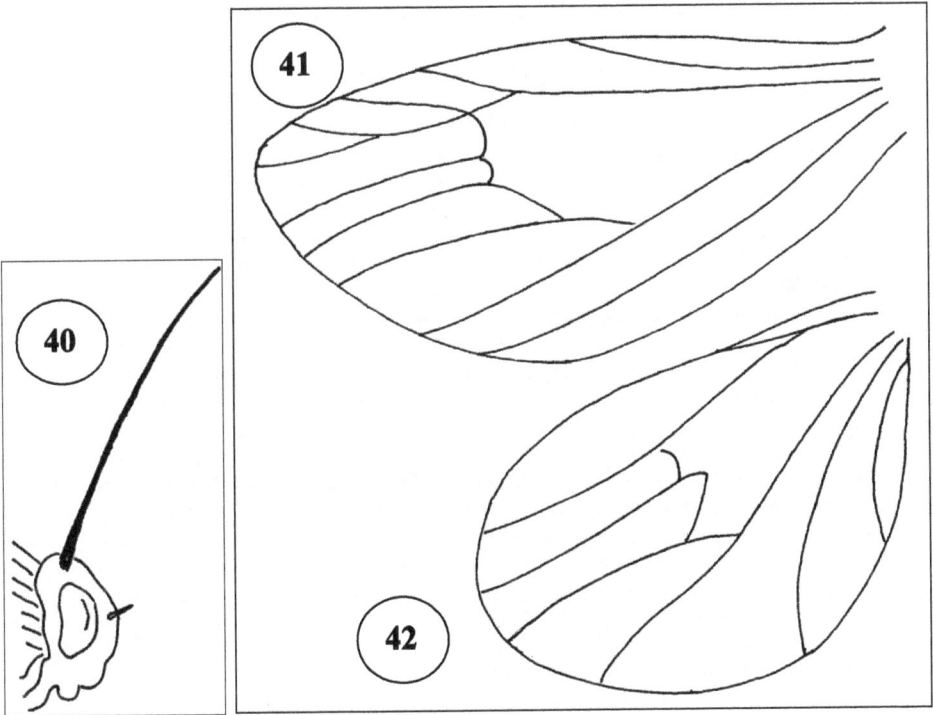

Figure 40: Head and Appendages.

Figure 41: Fore Wing Venation.

Figure 42 : Hind Wing Venation.

Host Plants

Ficus religiosa, Pakur, *Ficus* spp.

Holotype

Female, India, Maharashtra, Kumbharli Ghat, Koyanannagar, Coll. 3 VIII 2010, V. Y. Kadam, pinned insect in insect box, labeled as above.

Paratype

Female 3, Male 1, sex ratio (M:F) (1:3), coll. from July to October, 5 VIII 2009, Kolhapur, T. V. Sathe, Female 3, Coll. 3 IX 2009, Radhanagari.

Distribution

Western Ghats of Maharashtra (Kolhapur and Satara), throughout India.

Control

Spray carbaryl 0.15 per cent or Malathion 0.03 per cent.

Remarks

This species, runs close to *Argina syringa* Cramer (Hampson, 1976) by having following characters

1. Head, thorax and fore wing pale pinkish brown. However, it difffer from above species by having following characters.
2. Abdomen bright yellow.
3. Fore wing bright yellow.
4. Number and size of black spots on fore and hindwings.
5. Frons black.
6. Length proportion of tibia and femur.

Argina indica sp.nov. (Figure 43)

Female

Body 19 mm long, 5 mm broad, wing expanse 55 mm, antenna 7 mm long, Hind leg 11.2 mm long.

Head

2 mm long, 1.7 mm broad; eyes large and round; frons brownish, genae brownish, occular distance 1.4 mm, labrum flap like, concelled by clypeus, proboscis coiled. **Antenna** : 7 mm long, 0.4 mm broad, blackish brown, scape 0.5 mm long, pedicel 0.3 mm long, flagellum 6.2 mm long.

Thorax

5 mm long, 53 mm broad, brownish on dorsal side, yellow on ventral side, collar brownish, **Fore wings :** 26 mm long., 11 mm broad, brown in colour with white spots, margin serrate. **Hind wing** : 18 mm long 15 mm broad, yellowish with black spots.

PLATE 15—Figure 43: *Argina indica* **sp.nov.**

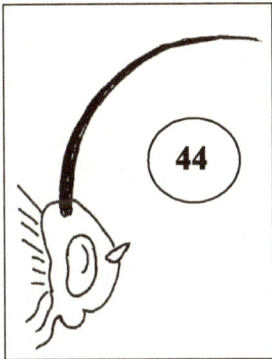

Figure 44: Head and Appendages.

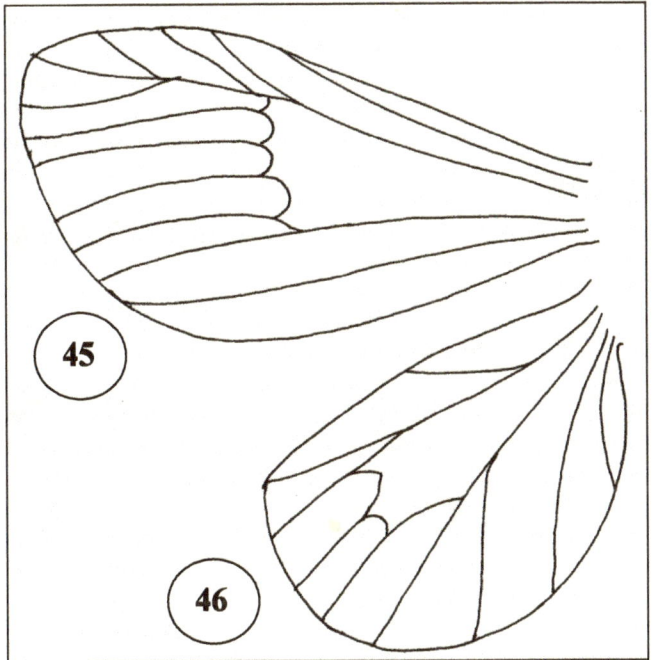

Figure 45: Fore Wing Venation.

Figure 46 : Hind Wing Venation.

Hind leg : 11.2 mm long, brown coloured, coxa 3 mm long, Trochanter 0.2 m long, femur 5 mm long, tibia 2 mm long, tarsus 1 mm long, claw reduced and curved.

Abdomen

12 mm long, 3mm broad, yellowish brown with dark black spots on lateral and dorsal side.

Male

Similar to female except sexual characters and smaller in size.

Host Plants

Ficus religiosa.

Holotype

Female, India, Radhangari, Kolhapur, coll. 27 VIII 2009, V.Y. Kadam, pinned insect in insect box and labeled as above.

Paratype

Female 4, male -2, sex ratio (M: F) 1:2, coll. July to November, T. V. Sathe.

Distribution

Throughout India, western Ghats of Maharashtra, Radhanagari, Plain region of Kolhapur and Satara.

Control

Spray carbaryl 0.15 per cent or Malathion 0.03 per cent.

Remarks

According to Hampson (1976) *Argina indica* sp.nov. runs close to *Argina argus* by having following characters.

1. Head, thorax and fore wing brownish red.
2. Two black spots on collar.

However, it differs from *A. argus* by having following characters.

1. Black spot size and number on hind wings.
2. Length proportion of femur to tibia.
3. Collar two black spots ringed with grey colour.
4. Abdomen colour lighter than hind wing.
5. Hind wing with 8 large black spots and 3 small black spots.
6. Four blacks spots on thorax.
7. Five blacks spots on costal line.

Argina guttata Hampson (Figure 47)

Female

Body 17.5 mm long, 2.5 mm broad, antenna 12 mm, hind leg 12.3 mm, wing expanse 44.5 mm.

Head

2 mm long, 1.6 mm broad; eyes small brown, occular distance 0.2 mm, frons browhish; antenna 12 mm, long, 0.3 mm broad, scape 0.4 mm, pedicel 0.3 mm, flagellum 11.3 mm.

Thorax

3. 5 mm long, 2.5 mm broad. **Fore wing :** 21 mm long, 10 mm broad, yellowish with many small black spots. **Hind wing :** 16 mm long, 11 mm broad, yellowish with large black spots. **Hing leg. :** 12.3 mm long, coxa 3.0 mm, trochanter 0.2 mm, femur 4.5 mm, tibia 2.1 mm, tarsus 2.5 mm.

Abdomen

12 mm long, 2 mm broad, yellowish dorsally and ventally.

Male

Similar to female except sexual characters and smaller in size.

Host Plants

Citrus limettia.

Holotype

Female, India, koyananagar, Satara, coll. 13 IX 2011, V. Y. Kadam, pinned insect in insect box and labeled as above.

Paratype

Female 2, Male 1, sex ratio (M:F) 1:2, coll. from July to October, V.Y. Kadam.

Distribution

Westerm Ghats of Maharashtra.

Control

Spray carbaryl 0.15 per cent or Malathion 0.03 per cent.

Remarks

According to the Hampson (1976) this species is *Argina guttata.* In the text some additional characters are given for this species since it is very poorly described by Hampson (1976).

PLATE 16—Figure 47: *Argina guttata* Hampson.

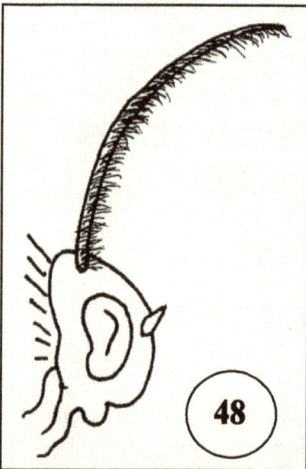

Figure 48: Head and Appendages.

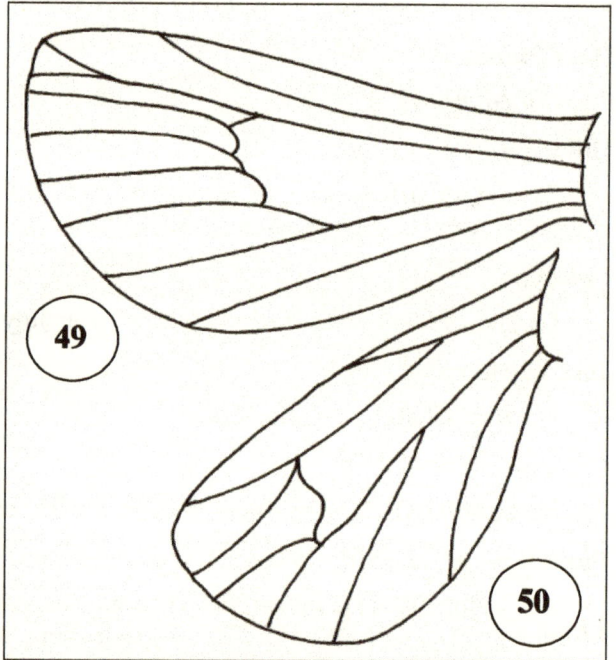

Figure 49: Fore Wing Venation.
Figure 50: Hind Wing Venation.

Argina satarensis sp.nov. (Figure 51)

Female

Body 19 mm long, 4 mm broad, antenna 11 mm long, hind leg – 12.8 mm long, wing expanse 44 mm.

Head

2 mm long, 1.5 mm broad;   eyes small, black and round, occular distance 0.3 mm; frons yellowish; antena 11 mm long, scape 0.3 mm, pedicel 0.4 mm, flagellum 10.3 mm.

Thorax

5 mm long, 4 mm broad, wing expanse 44 mm   **Fore wing :** 20 mm long, 9 mm broad, yellowish with black spots. **Hind wings:** 16 mm long, 7 mm broad, yellowish with black spot. **Hind leg:** 12.8 mm long, coxa 4.0 mm, trochanter 0.2 mm, femur 4.1 mm, tibia 2 mm, tarsus 2.5 mm.

Abdomen

12 mm long, 2mm broad; dorsally brownish ventrally yellowish and black spots are on lateral side.

Host Plants

Tarmarind.

Holotype

Female, India, Panchgani, Satara, coll. 27 VIII 2010, V. Y. Kadam, pinned insect in insect box and labeled as above.

Paratype

Female 4, male 2, sex ratio (M:F) 1:2, coll. from Radhanagari, Coll. July to December, V. Y. Kadam.

Distribution

Western Ghtats of Maharashtra, Myanmar.

Control

Spray carbaryl 0.15 per cent or Malathion 0.03 per cent.

Remarks

This species runs close to *Argina purdalina* Hampson by having following features.

1. Head, Thorax and fore wing not dark brown with pale yellow.

However, it differs from above species by having following features.

1. Basal region of fore wing form circle of six black spots.
2. Abdomm yellow dorsally and ventrally.
3. Length portion of femur to tibia.
4. General body colour and wing spots.

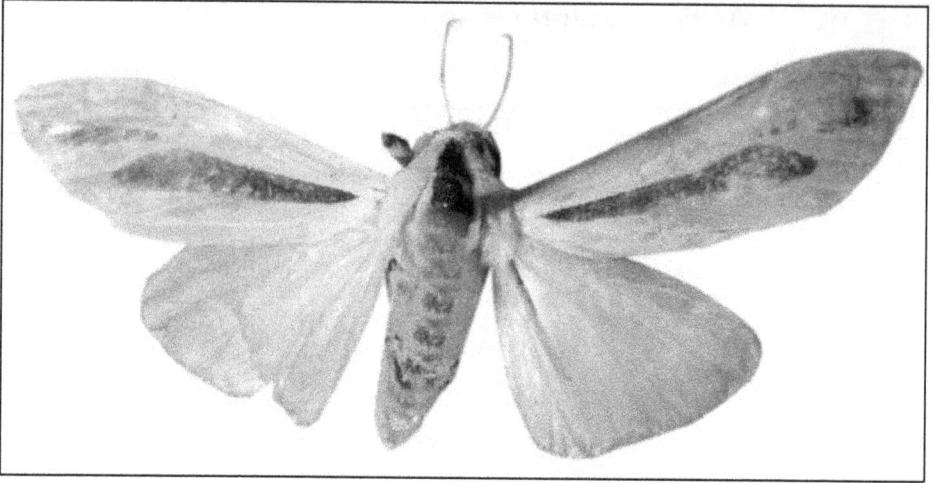

PLATE 17—Figure 51: *Argina satarensis* **sp.nov.**

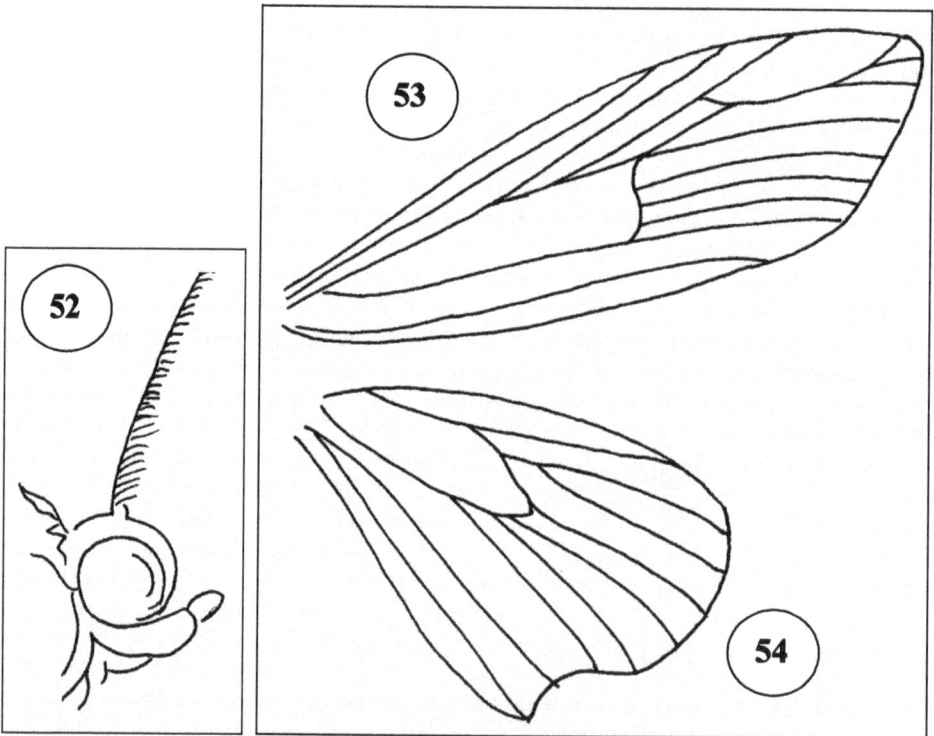

Figure 52: Head and Appendages.

Figure 53: Fore Wing Venation.
Figure 54: Hind Wing Venation.

SUB FAMILY ARCTIINAE

The subfamily Arctiinae is characterised by having

1. Proboscis often absent or very minute.
2. The reticulum of the male consists of a strong bar from the costal nervure of fore wing instead of the usual tuft of hair.
3. The legs are usually smooth and the spur minute.
4. Thick builded Ermines and tiger moths are grouped under this subfamily. Larvae with four pairs of fore legs and clothed with very long hair. Larva form cocoon with hair for pupation.

KEY TO THE GENERA

1. Fore wings with the outer margin some what erect   *Spilosoma*
2. Palpi extending far beyond the frons   *Areas*
3. Palpi extending beyond the frons   *Creatonotus*
4. Hind wings with vein 6 and 7 from cell   *Moorea*
5. Fore wing narrow   *Pelochyta*

GENUS – *SPILOSOMA*

Spilosoma, Steph III Brit Ent., Haust ii P. 74 (1829)

Alpenus, wlk. cat iii. P. 686 (1855)

Spilaretia, Butl, Cist Ent. ii. P. 39 (1875)

Challa, Moore, PZS. 1979, P. 398.

Type, *S. lubricipedium*, Linn., from Europe.

Range : Nearctic, Palaearctic and oriental regions

The genus *Spilosoma* is characteristed by having

1. Palpi short, porrect and frigned with hair
2. Antenna bipectinate in male, serrate in female.
3. Mid tibia with terminal pair of minute spurs and hind tibiae with two pairs.
4. Fore wings long and narrow, veins 3, 4, 5, from angle of cell, 6 from upper angle, 7, 8, 9, 10 stalked.
5. Hind wings with veins 3, 4, 5 from angle of cell, 6 and 7 from upper angle, 8 from middle cell.

KEY TO THE SPECIES OF THE GENUS SPILOSOMA

1. Abdomen yellow in both sexes   3
   Abdomen crimson inboth sexes   2

2. Male with branches of antennae long   4

    Antenna whitish   *S. strigatulum*

    Antenna not tipped with white   *C. motanum*

    Antenna blackish   *S. sathei*

Spilosoma sathei sp.nov. (Figure 55)

Male

Body 20 mm long, 5 mm broad, pinkish dorsally, antenna bipectinate, 7 mm long, blackish grey coloured not tiped with white, wing expanse 44 mm.

Head

2 mm long, 2 mm broad, blackish; eyes large round, occular distance 1.0 mm, antenna bipectinate 7 mm long, scape 0.6 mm, pedicel 0.4 mm, flagellum 6 mm, grey coloured.

Thorax

6 mm long, 5 mm broad, dorsally light grey, ventrally orange coloured. Four black spots on throrax. **Fore wing**: 20 mm long, 11 mm broad, whitish with black spots, serrate margin costal line with blackish stripe and blackish spots, series of black spots on the marginal line of fore wing. **Hind wing** : 16 mm long, 11 mm broad, dorsally light orange and ventrally yellowish. **Hind leg** : 11.7 mm long, coxa 3 mm, trochanter 0.2 mm, femur 4 mm, tibia 2 mm, tarsus 2.5 mm, claws curved and pointed at the tip.

Abdomen

13 mm long, 5 mm broad, crimson in male and female, pinkish with black spots laterally and ventrally orange coloured with black spots. Ventral mid portion without spots.

Female

Similar to male except sexual characters and larger in size.

Host Plants

Fagiolus sp.

Holotype

Male, India, Maharashtra, Shahuwadi, coll. 7 VIII 2007, V. Y. Kadam, pinned insect in insect box and labeled as above.

Paratype

Female 4, male 2, sex ratio (M.F.) 1:2, coll. from July to November, Kolhapur, T. V. Sathe.

PLATE 18—Figure 55: *Spilosoma sathei* **sp.nov.**

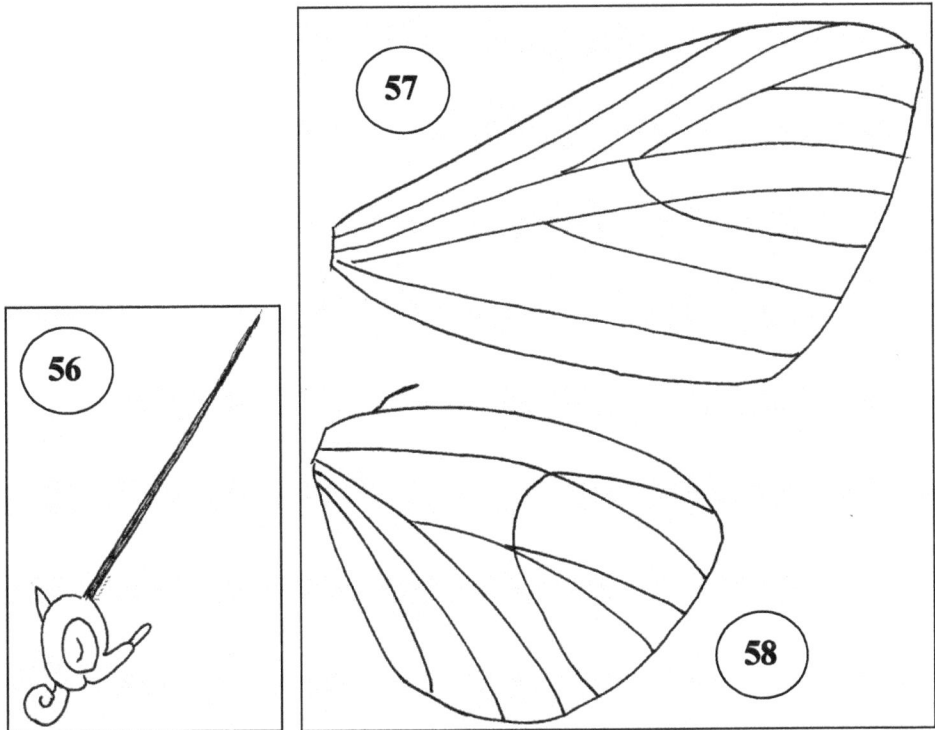

Figure 56: Head and Appendages.

Figure 57: Fore Wing Venation.
Figure 58: Hind Wing Venation.

Distribution

Western Ghats of Maharashtra, Kolhapur, Satara.

Control

Spray carbaryl 0.15 per cent or Malathion 0.03 per cent or Rogor 0.03 per cent.

Remarks

Accordingly Hampson (1976). *Spilosoma sathei* sp. nov. runs close to *Spilosoma strigatulum* Walker by having following characters

1. Male antenna with long branches.
2. Abdomen crimson in male and females.

However, it differs from above species by having following characters.

1. Antennal branches are not white but blackish grey.
2. Costal line of fore wing with black stripe and black spot.
3. Series of small marginal black spots on fore wing.
4. Four black spots on thorax.
5. Hind tibia with minute and with two pair of spurs.

GENUS *CREATONOTUS*

Creatonotus, Hiibn verz. p. 169 (19118)

Rhodogastria, Hiibn Verz. p. 172.

Aloa, Wlk Cat iii, p. 699 (1855).

Savara Wlk. Cat. XXXi, P 320 (1864).

Bucaea Wlk Cat. Xxxv p. 1983 (1866).

Type, C wlk interruptus, Linn.

Range – Throughout India, Sri Lanka, Myanmar,

China, Penag, Java.

The genus *Creatonotus* is characterised by

1. Palpi short and porect.
2. Hind tibia with one pair of spurs.
3. Fore wing narrow, veination as in *Spilosoma*.

Fore wings in some specimens with vein10 from cell, and vein 5 in both wings.

Sometimes from above angle of cell.

11 species have been reported from this genus from India.

Creatonotus gangis (Linnaeus) (Figure 59)

Creatonotus interruptus Gmel syst No. 2533

Creatonotus continautus Moore AMNH 1877

Bombyx francisea Fabr. Mant. Ins. pp. 131.

Male

Body 15.5 mm long, 1.6 mm broad colourless, antenna 13 mm long,

hind leg 13. 0 mm, wing expanse 41 mm.

Head

2.3 mm long, 1.7 mm broad, pale pinkish ochreous; eyes small black, occular distance 0.2 mm; frons brownish; antenna 13 mm long, scape 0.6 mm, pedicel 0.5 mm, flagellum 11.9 mm.

Thorax

4.2 mm long, 3 mm broad, pale pinkish ochreous. **Fore wing :** 19 mm long, 8 mm broad, colourless with broad black streak at middle supplemented by another small black streak. **Hind wing :** 12 mm long, 10 mm broad, colourless, pale fuscous, submarginal series of black spots. **Hind leg. :** 13.0 mm long with 1 pair of spurs, coxa 5.0 mm, trochanter 0.3mm, femur 4.2 mm, tibia 1.5 mm, tarsus 2.0 mm.

Abdomen

9 mm long, 3 mm broad, crimson orange dorsally with black spots and brownish ventrally.

Male

Similar to female except sexual characters.

Host Plants

Pomegranate, *Punica grantum* Linnaeus,Wild Banana, *Musa* sp.

Holotype

Male, India, Maharashtra, Radhanagari, coll. 17 VIII 2008, V. Y. Kadam, pinned insect in insect box and labeled as above.

Paratype

Female 1, male 2, sex ratio (M :F) 2:1, coll. July to December, V. Y. Kadam.

Distribution

Andrha Pradesh, Maharashtra, western Ghats of Maharashtra, Plains of Kolhapur and Satara, out side India Shri Lanka and Myanmar.

Control

Spray carbaryl 0.15 per cent or Malathion 0.03 per cent or Rogor 0.03 per cent.

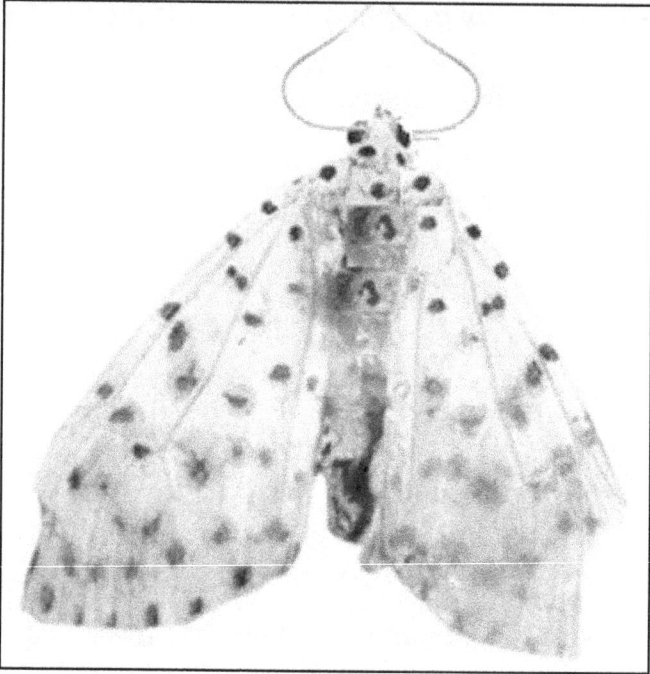

PLATE 19—Figure 59: *Creatonotus gangis* (Linnaeus).

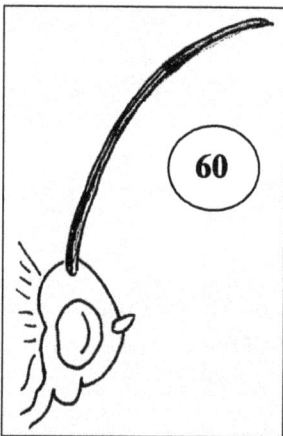

Figure 60: Head and Appendages.

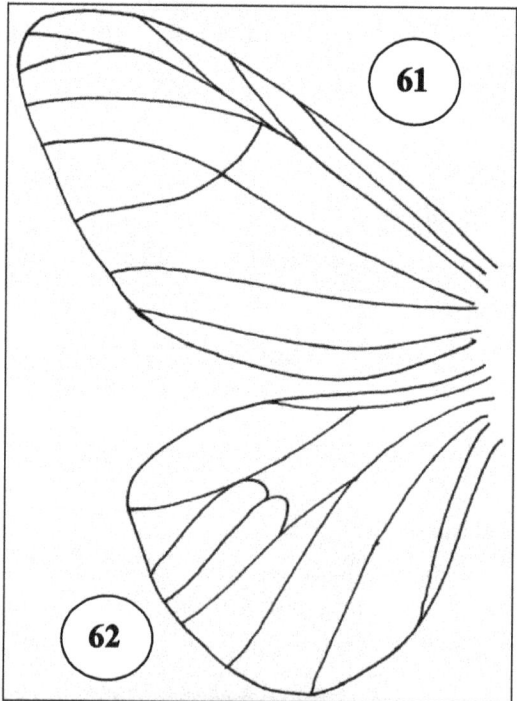

Figure 61: Fore Wing Venation.
Figure 62: Hind Wing Venation.

Remarks

According to Hampson (1976) this species runs close to *Creatonotus interruptus* Gmel by many characters.

GENUS *PELOCHYTA*

Pelochyta, Hubn Vorz P. 171 (1818)

Amerila, Wlk. Cat. Iii, P. 725(1855)

Type, *P. vidua,* Cram, from Africa.

Range: Africa, Formosa, throughout India, Myanmar and Shri Lanka, Australia, New Britain.

The genus *Pelochyta* Hubn. is characterised by having

1. Palpi porrectly upturned and slender.
2. Antenna nearly simple, fore wing with costa arched, the outer margin long and obliqualy curved, veins 3,4,5. from angle of cell, 6 from below upper angle, 7,8,9 stalked.10 from cell.
3. Hind wing with the anal angle slightly produced, outer margin nearly straight, apex slightly acute veins 3,4,5 from close to angle of cell, 6 and 7 from upper angle, 8 from near end of cell.

From this genus only single species, *Pelochyta astrea* Drury has been reported and described from India.

KEY TO THE SPECIES OF THE GENUS *PELOCHYTA*

1. Fore and hind wing hyaline. Two black spots at basal region of fore wing   *astrea*
2. Three black spots at basal region of fore wing   *sathei.* sp. nov.

Pelochyta sathei sp.nov. (Figure 63)

Female

Body 21.5 mm long, 4.5 mm broad, antenna 16 mm, hind leg 17 mm long, wing expanse 70 mm.

Head

2.5 mm long, 2 mm broad; eyes browinish round, occular distance 0.3 mm; frons cream coloured; palpi crimson red, antenna reddish brown, 16 mm long, scape 0.7 mm, pedicel reddish brown 0.5 mm, flagellum 14.8 mm. Head with two black spots, Collar with 4 black spots dorsally 2, ventrolaterally 2.

Thorax

5 mm long, 4.5 mm broad, pro, meso and metathorax with a pair of black spots, wing expanse 70 mm. **Fore wing:** 33 mm long, 11 mm broad, hyaline, grey coloured with 3 black spots at basal region. Large pale fuscus patch on apical area and costal

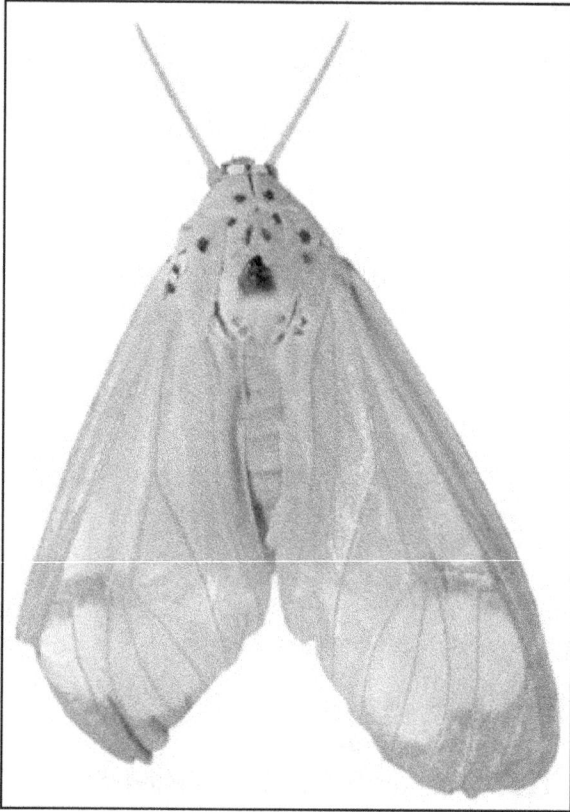

PLATE 20—Figure 63: *Pelochyta sathei* **sp.nov.**

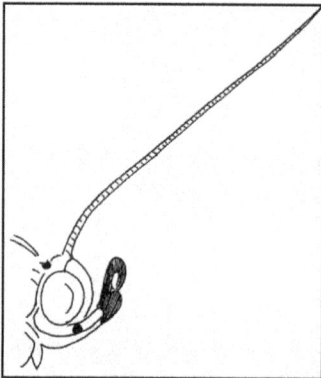

Figure 64: Head and Appendages.

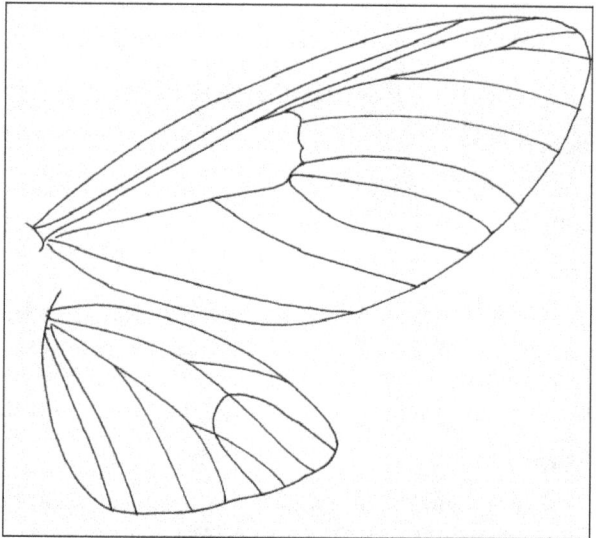

Figure 65: Fore Wing Venation.
Figure 66: Hind Wing Venation.

margin. **Hind wing :** 22 mm long, 12 mm broad, hyaline, light grey, with marginal small fuscus band. **Hind leg :** crimson, hyaline, 17 mm long, coxa 4.7 mm, trochanter 0.3 mm, femur 5.5 mm, tibia 3.5 mm, tarsus 3 mm.

Abdomen

14 mm long, 4 mm broad, orange dorsally and ventrally, cream coloured with black spots laterally. Out of which one row on dorsal side which is crimson and anterior row of small black spots on ventral side.

Host Plants

M. indica

Holotype

Female, India, Maharashtra, Kolhapur, Anuskura Ghat, coll. 7 VII 2010, V. Y. Kadam.

Paratype

Female 3, male 3, sex ratio (M:F) 1:1, coll. June to November, V. Y. Kadam, pinned insect preserved in insect box and labeled as above.

Distribution

Western Ghats of Maharashtra, Kolhapur, Satara.

Control

Spray carbaryl 0.15 per cent or chloropyriphos 0.02 per cent.

Remarks

According to the Hampson (1976) this species runs close to *Pelochyta astrea* Durry by having following characters

1. Fore wing hyaline, and
2. Hind wing hyaline.

However, it differ from above species by having following characters

1. Size of pale fuscus patches on fore wing and hind wing.
2. 3 black spots at basal region of fore wing.
3. Legs with outer side crimson on coxa, femur and tibia.

GENUS *MOOREA*

Grotea, Moore P.Z.S 1865,P797 (Praeocc)

Type : *Moorea argus.*

Range : Sikhim, Khasi Hills.

The genus *Moorea* was ercted by Hampson in 1897.It is characterised by

1. Palpi porrect with moderate length, first two joints fringed with hair.

2. Antennae simple.

3. Hind tibia with two pairs of minute spurs.

4. Fore wing with vein 3 from before end of cell. 4 and 5 from the end, 6 from upper angle, 7,8,9,10 Stalked,11 running close along side of or touching 10.

5. Hind wing with vein 3 from before angle of cell, 4 and 5 from angle 6 and 7 from upper angle and 8 from below middle of cell.

From India only one species has been reported. In the text one more new species has been reported and described.

KEY TO THE SPECIES OF GENUS *MOOREA*

1. Head, thorax,golden yellow. Two black spots on collar   4, 5

2. Head thorax two black spots collar   3.5

3. Head, thorax brown.Thorax with 3 black spots   *indica* sp.nov.

4. Head, thorax brownish yellow.Hind wing pale yellowish,
    large black patches   *marathi* sp.nov.

5. Fore wing golden yellow with bluish white spots, thorax
    with two black spots   *M. argus*

Moorea indica sp.nov. (Figure 67)

Female

Body 23 mm long,5 mm broad, antenna simple, 9 mm, wing expanse 55 mm,hind leg 13.8 mm.

Head

2 mm long, 1.5 mm broad,dark brown; eyes rounded, occular distance 0.5 mm; antenna simple, brownish, 9 mm long, scape 0.3 mm, pedicel 0.4 mm, flagellum 8.3 mm.

Thorax

6 mm long,5 mm broad, dark brown, 4 black spots on thorax, wing expanse 55 mm. **Fore wing** : 25mm long, 9 mm broad,dark brown black spots, serrate margin, transversely arranged series of black spots. **Hind wing :**14 mm long, 12 mm broad, golden yellow, black large patch at anal margin, 'V' shaped black patch at central part, apical margin with large black patch. **Hind leg:** 13.8 mm long, coxa 5 mm orange with black spots, trochanter 0.3 mm, femur 5 mm, tibia 1.5 mm, tarsus 2 mm.

Abdomen

15 mm long,4 mm broad, dorsally orange golden yellow with transverse black patches,ventrally brown. Abdominal last segment with large black patch dorsoventrally.

PLATE 21—Figure 67: *Moorea indica* sp.nov.

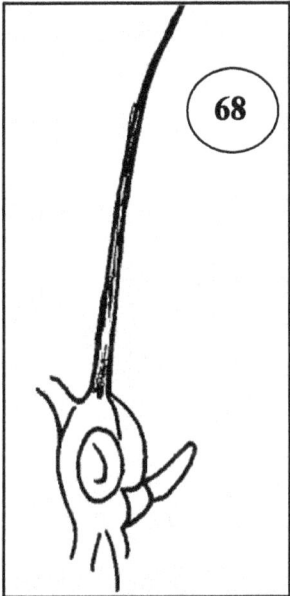

Figure 68: Head and Appendages.

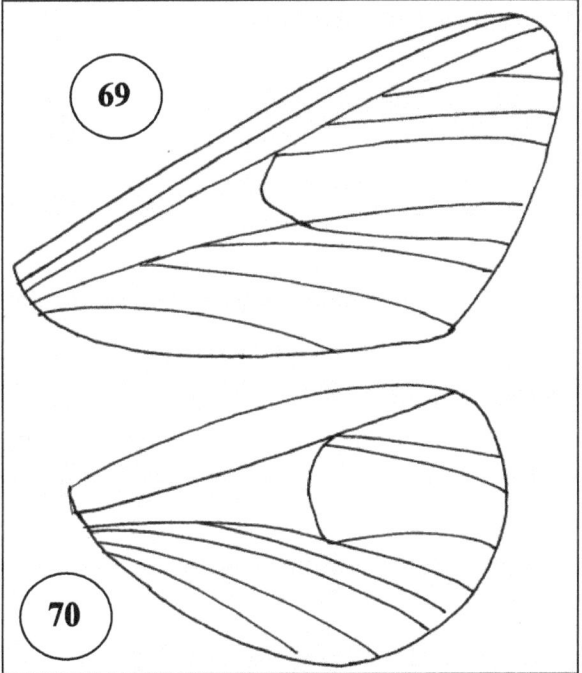

Figure 69: Fore Wing Venation.
Figure 70: Hind Wing Venation.

Male

Similar to female except sexual characters.

Host Plants

Musa acuminata

Holotype

Female, India, Maharashtra, western Ghats, coll.16 VIII 2011, T.V.Sathe, pinned insect in insect box and labeled as above.

Paratype

Female 2, Male 1, sex ratio (M: F) 1: 2, coll. from June to Dec. Panchgani,V.Y. Kadam.

Distribution

Western Ghats of Maharashtra, Kolhapur and Satara, plains of Satara and Kolhapur.

Control

Spray carbaryl 0.15 per cent or chloropyriphos 0.02 per cent.

Remarks

According to the Hamposn (1976) this species runs close to *Moorea argus* by having following characters:

1. Collar with two black spots.
2. Abdomen with golden yellow colour.

However, it differs from the above species by having follwings characters.

1. Thorax with 4 black spots.
2. Thorax and head dark brown.
3. Number and size of black patches on fore wing.
4. Number and size of black patches on hind wing.

Moorea marathi sp.nov. (Figure 71)

Female

Body 20mm long, 4 mm broad, antenna 11mm, simple, hind leg 13.4 mm, wing expanase 56 mm.

Head

2 mm long, 1.5 mm broad,brownish yellow; eyes large black, occular distance 0.5 mm; frons grey. **Antenna:** Simple, 11mm long, scape 0.3 mm, pedicel 0.4 mm, flagellum 10.3 mm.

Thorax

5 mm long, 4 mm broad, brownish yellow. **Fore wing:** 26 mm long,12 mm broad, grey with dark brown spots. **Hind wing:** 19 mm long,15 mm broad, pale yellow with

PLATE 22—Figure 71: *Moorea marathi* **sp.nov.**

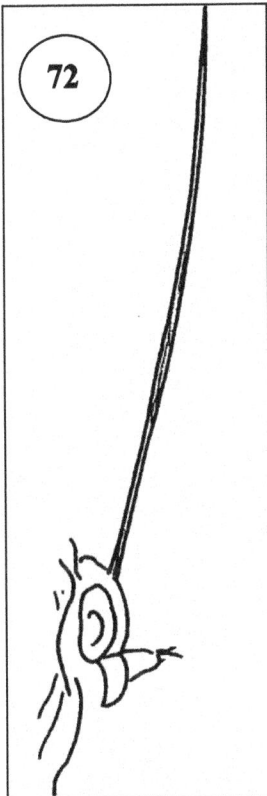

Figure 72: Head and Appendages.

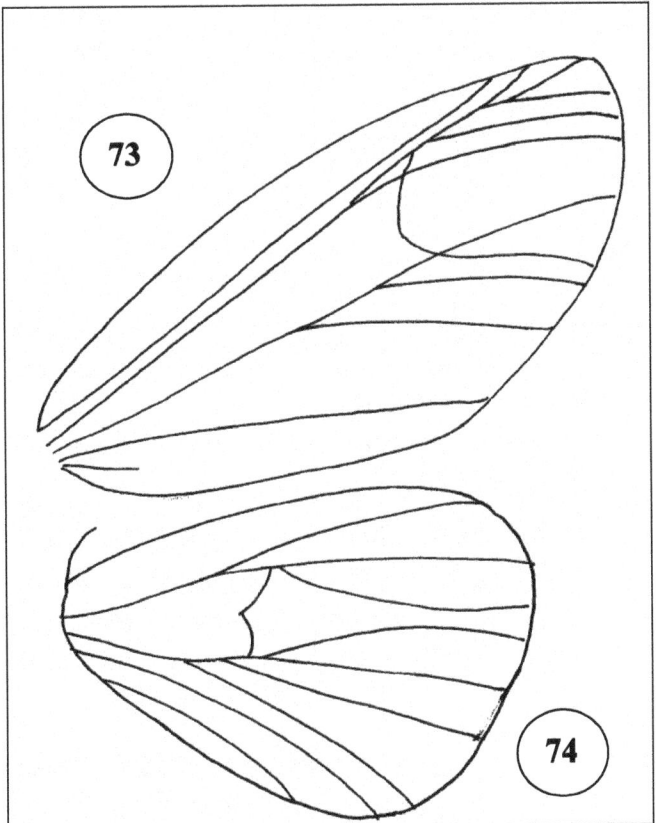

Figure 73: Fore Wing Venation.

Figure 74: Hind Wing Venation.

dark brown patches in four rows in slightly zig-zag manner, second marginal row with large black patches. **Hind leg :** 13.4 mm long, coxa 4 mm, trochanter 0.3 mm, femur 4.5 mm, tibia -2.1 mm, tarsus 2.5 mm.

Abdomen

13 mm long, 3.4 mm broad, light yellow with transverse black bands.Tip of abdomen with black patch.

Male

Similar to female except sexual characters.

Host Plants

Musa acuminata

Holotype

Female, India, western Ghats of Maharashtra, Koyananagar, coll.16VIII 2011, T.V.Sathe,pinned insect in insect box and labeled as above.

Paratype

Female 2, male 1,sex ratio (M:F)1:2, coll. from June to Dec, Panchgani, V.Y.Kadam.

Distribution

Western Ghats of Maharashtra, Kolhapur and Satara, plains of Satara and Kolhapur.

Control

Spray carbaryl 0.15 per cent or chloropyriphos 0.02 per cent.

Remarks

According to the Hampson (1976) this species runs close to *Moorea argus* by having follwing characters.

1. Collar with two black spots ringed with yellow colour.
2. Antennae simple.

However, it differs from the above species by having following characters,

1. Abdomen pale yellow with transverse black bands.
2. Fore wing black spots by size and number.
3. Size and number of black spots on hind wing.
4. Veination of fore and hind wings.

FAMILY : SPHINGIDAE

The family Sphingidae is charactered by

1. Antennae being gradually thickened into a club, which is pointed at the tip, nearly always hooked, with a small tuft of hair at the extremity, in the males of nearly all the genara there are band of cilia on the undersurface.

2. Palpi very thickly scaled, the third joint minute and buried in the scales.

3. Leg strong with well developed spurs and spined tarsi.

4. Fore wing elongated and narrow, the subcostal is very close to the costa, vein 1 forked at base.

5. Hind wing small, costal nervure arising free, with a bar between it and subcostals, two internal veins. The members are diurnal or crepuscular and very powerfull fliers. The shape and colouration are eminently variable, larvae are smooth, nearly always with a horn on 11th segment.

KEY TO THE SUBFAMILIES

a. The proboscis very short and thick | Acherontiinae

b. The proboscis very short and slight | Simerinthinae

c. Imago : The proboscis of moderate length.

The apex of fore wing much produced, male with small lateral expansions to abdomen | Ambulycinae

Apex of fore wings but slightly produced, male with small lateral expansions to abdomen | Chaerocamoinae

d. The probose is very long, abdomen conical in both sexes | Sphinginae

e. Abdomen with a medial pair of lateral tuft of hair on last segment more or less in both sexes | Macroglossinae

SUB FAMILY - MACROGLOSSINAE

The family Macroglossinae is characterised by having one medial and two lateral tuft of hair at end of abdomen.

KEY TO THE LOCAL GENERA

1. Fore wing more then twice length of antennae. Antennae abruptly thickening to the club. Cell of hind wing normal | Rhopalopsyche

2. Fore wing with vein 6 given off after end of cell. Hind wing with veins 3, 4, and 6, 7 stalked the cell extremely short | Cephonodes

GENUS : RHOPALOPSYCHE

Rahopalopsyche, Bull. PZS 1875,

Type, *R. nycteris* Koll.

Range: Himalayas and peninsular India.

The genus *Rhophalopsyche* is characterised by

1. Antennae clubbed, the basal half thin, the hook short and slight with no bands of cilia in male.

2. Palpi with the apex porrect and acutely scaled.

3. Fore wing not more than twice the length of antennae, the outer margin excurved.

4. Flight diurenal

From India two species have been reported and described (Hampson, 1976)

KEY TO THE SPECIES OF RHOPALOPSYCHE.

Head thorax and abdomen grey brown. Hind wing with bright yellow baselly   *nycteris*

Hind wing dark brown red basally   *bifaciata*

Rhopalopsyche nycteris Koll. (Figure 75)

Female

Body 29 mm long. 4 mm broad,wing expanse 52 mm, antenna 11 mm long, 0.3 mm broad, hind leg 22.3 mm long.

Head

5 mm long, 3 mm broad, grey brown in colour; eyes large oval; frons blackish grey, ocular distance 1.4 mm; antenna 11 mm long, 0.4 mm broad, grey coloured; scape 0.6 mm, pedicel 0.5 mm, flagellum 9.9 mm, antennae clubbed the basel half thin, the hook short and slight.

Thorax

7 mm long, 10 mm broad. **Fore wing:** 23 mm long, 7 mm broad, grey brown, some subbasal indistinct lines, antemedial band reccurved toward the margin, three postmedial curved lines, square brown spot on the costa before the apex with a black spot below it, from which a waved obliqua line runs to the apex. **Hind wing:** 15mm long, 9 mm broad, black brown with a broad medial yellow band. **Hind leg:** 22.3 mm long, coxa 4 mm, trochanter 0.3 mm, femur 5 mm, tibia 9 mm, tarsus 4 mm, claws curved and pointed at tip.

Abdomen

17 mm long, 9 mm broad, dark grey dorsally and brown ventrally, 3 yellow bands/spots laterally, tuft of blackish hairs at tip of abdomen.

Male

Similar to female except sexual characters and antennae without band of cilia in male.

Host Plant

Vitis vinifera Linn.

Holotype

Female,India, Maharashtra, Yevteshwar, 17 IX 2010, V.Y. Kadam, pinned insect in insect box, labeled as above.

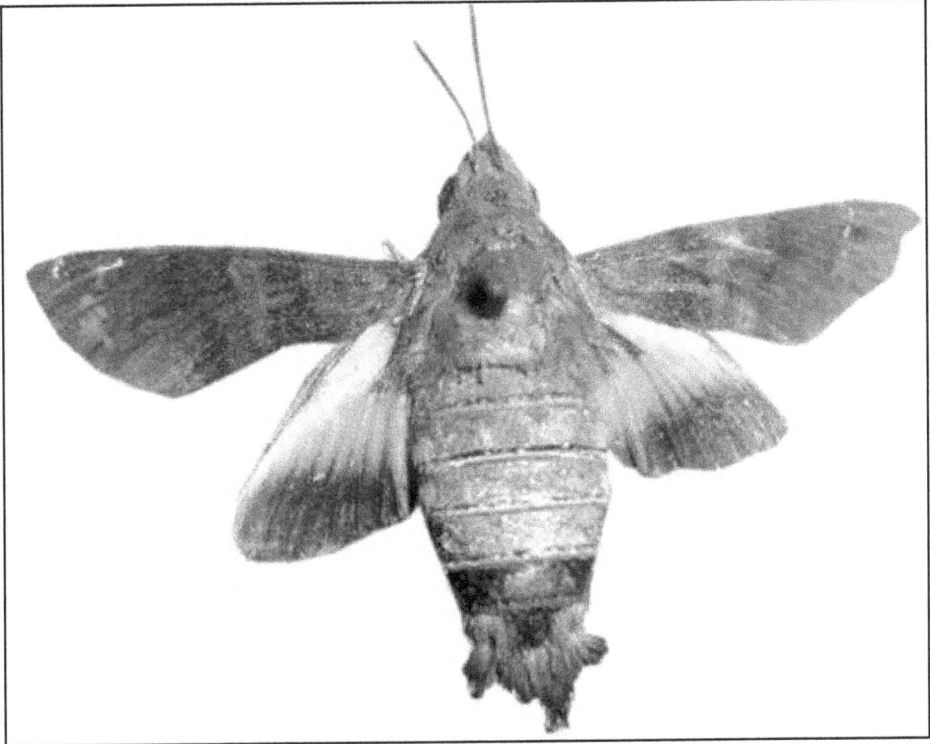

PLATE 23—Figure 75: *Rhopalopsyche nycteris* **Koll.**

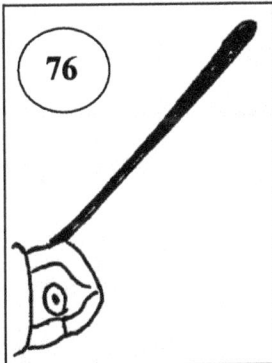

Figure 76: Head and Appendages.

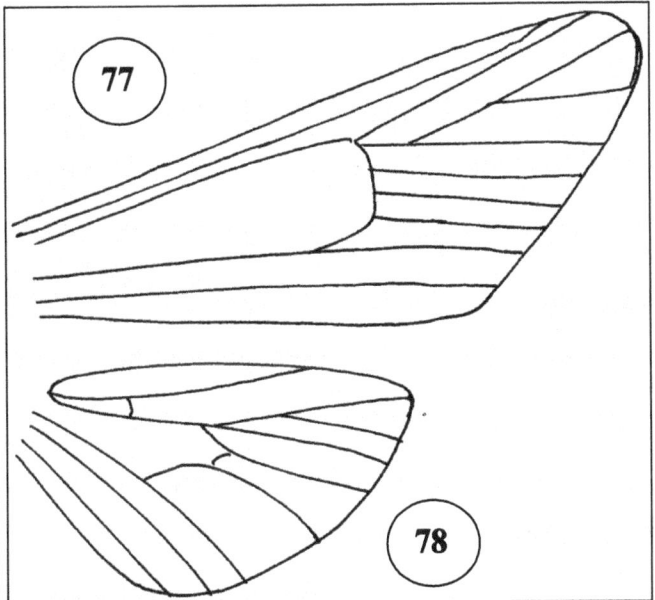

Figure 77: Fore Wing Venation.

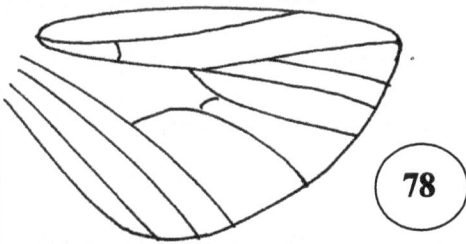

Figure 78: Hind Wing Venation.

Paratype

Female 4, Male -1 sex ratio (M:F) (1:4), coll. Maharashtra, Satara, Koyananagar from June to September, T. V. Sathe.

Distribution

Western Ghats of Maharashtra, Kolhapur, Satara plain region of Kolhapur.

Control

Spray Rogor 0.03 per cent or chloropyriphos 0.02 per cent, Malathion 0.03 per cent

Remarks

According to Hampson (1976) this species is *Rhopalopsyche nycteris* Koll. However, the description is poor hence redescribed in the text.

Rhopalopsyche bifaciata Butt. (Figure 79)

Female

Body 24 mm long, 9 mm broad, antenna 8 mm long, hind leg 17.3 mm, wing expanse 58 mm.

Head

4 mm long, 3 mm broad; eyes large round and black; occular distance 1.2 mm; frons blackish; antenna 8 mm long, scape 0.3 mm, pedicel 0.4, flagellum 7.3 mm, antenne being thickened in to a club.

Thorax

12 mm long, 9 mm broad, wing expanse 58 mm. **Fore wing :** 23 mm long, 11mm broad, blackish some subbasal indistinct lines, an antemedial band darker. The interspace between the first two postmedial lines filled in with dark so as to form band. **Hind wing :** 13 mm long, 10 mm broad, brownish, with band bright orange/ dark orange (on lateral bands of abdomen). **Hind leg :** 17.3 mm longs, with well developed spur and spined tarsi, coxa 4 mm, trochanter 0.3 mm, femur 6 mm, tibia 4 mm, tarus 3 mm.

Abdomen

15 mm long, 9 mm broad, dorsally black and ventrally brownish, Abdomen is having one medial black spot.

Male

Similar to female except sexual characters.

Host Plants

Morus nigra

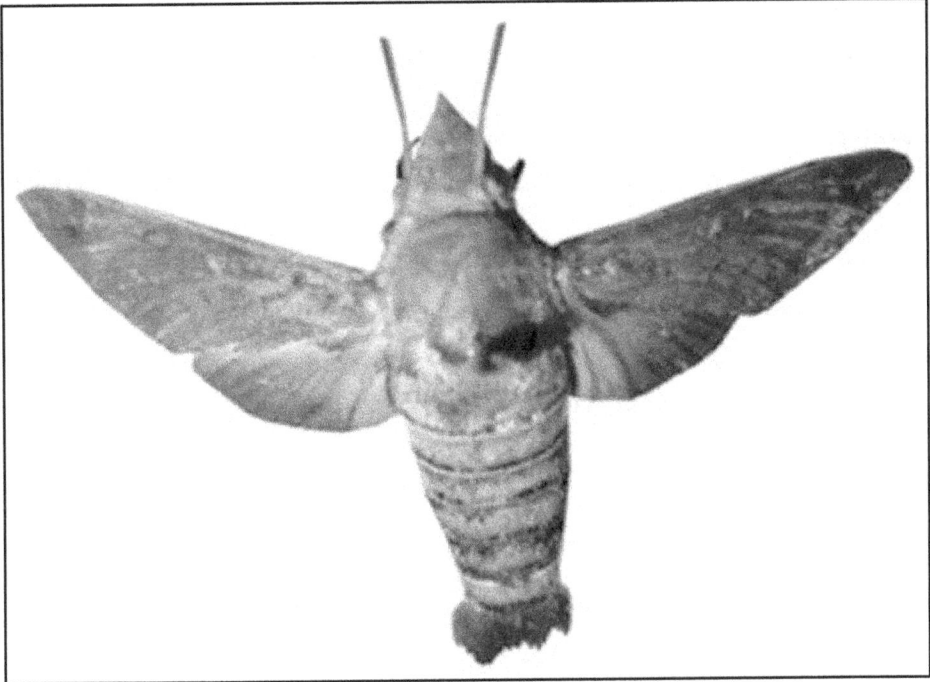

PLATE 24—Figure 79: *Rhopalopsyche bifaciata* **Butt.**

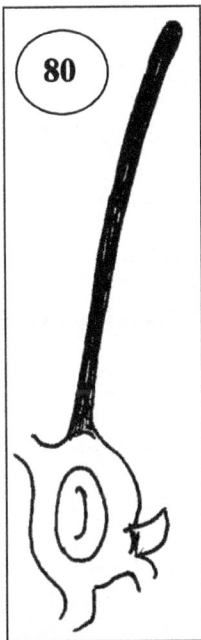

Figure 80: Head and Appendages.

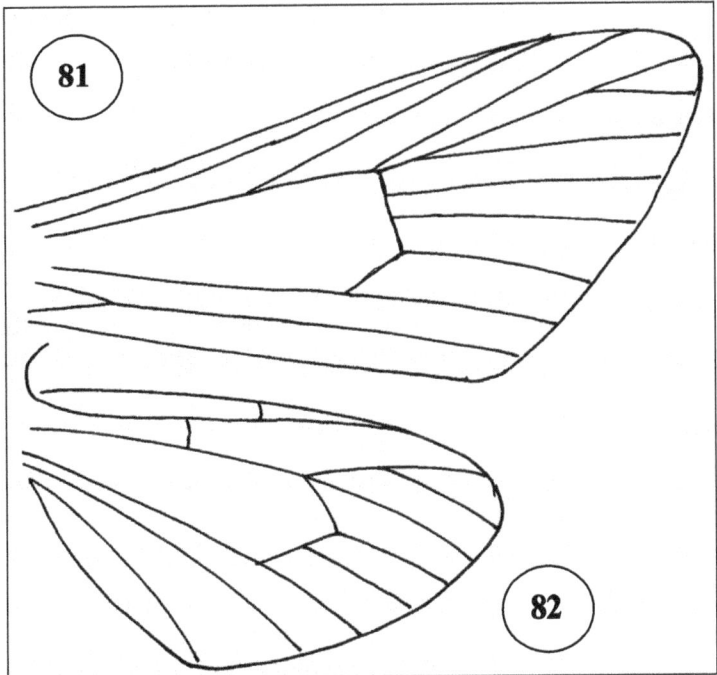

Figure 81: Fore Wing Venation.

Figure 82: Hind Wing Venation.

Holotype

Female India, Maharashtra. Shahuwadi, coll. 5 VIII 2011, V.Y. Kadam, pinned insect in insect box, labeled as above.

Paratype

Female 2, male 1, Sex ratio (M:F) 1:2, coll. from July to November, Radhanagari, T. V. Sathe.

Distribution

Western Ghats of Maharashtra, Kolhapur, Satara and plain region of Kolhapur.

Control

Spray Rogor 0.03 per cent or chloropyriphos 0.02 per cent, Malathion 0.03 per cent.

Remarks

According to Hampson (1976) this species is *Rhopalopsyche bifaciata* Butt.

However, poorly described by the prevoius author. Hence, redescribed and reported from western Ghats of Kolhapur and Satara region.

GENUS – *CEPHANODES*

Fore wing with vein 6 given off after end of cell.

Hind wing with veins 3, 4 and 6, 7 straight the cell extremely short.

Cephanodes indica sp.nov. (Figure 83)

Female

Body 30 mm long, 7 mm broad, antenna 12 mm, hind leg 20.4 mm, wing expanse 61 mm.

Head

3 mm long, 2 mm broad, grey brown; eyes large black, occular distance 0.5 mm; frons brownish, collar whitish grey, without black spots. **Antenna :** 12 mm long, scape 0.5 mm, pedicel 0.6 mm, flagellum 10.9 mm, being gradually thickened into club.

Thorax

8 mm long, 7 mm broad, grey brown/blackish brown, wing expanse 61 mm.

Fore wing: 27 mm long, 9 mm broad, hyaline, colourless, cell short, pterostigma at anterolateral tip of fore wing, costal line dark fuscus, anal half region dark fuscus.

Hind wing: 12 mm long, 7 mm broad, colourless, hyaline, basal half costal line dark fuscus. **Hind leg:** 24 mm long, coxa 6 mm, trochanter 0.4 mm, femur 5 mm, tibia 5 mm, tarsus 4 mm, with spur and spined tarsi.

Abdomen

19 mm long, 6 mm broad, black anal tuft present, two white spots ventrally.

PLATE 25—Figure 83: *Cephanodes indica* sp.nov.

Figure 84: Head and Appendages.

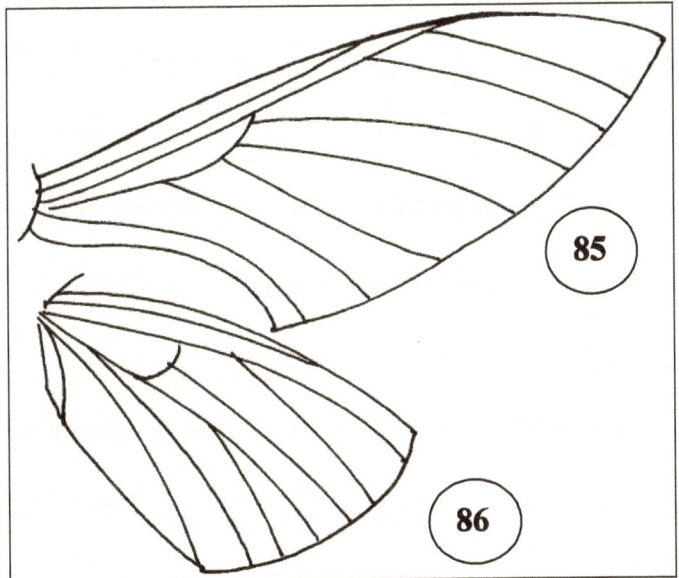

Figure 85: Fore Wing Venation.

Figure 86: Hind Wing Venation.

Male

Similar to female except sexual characters.

Host Plants

Terminalia bellirica

Holotype

Female, India, Maharashtra, Koyananagar, coll. 3 IX 2011, V. Y. Kadam, pinned insect in insect box and labeled as above.

Paratype

Female 3, male 2, sex rato (M:F) 1:1.5, coll. Shahuwadi, from June to November, T. V. Sathe.

Distribution

Western Ghats Kolhapur and Satara, plain region of Kolhapur.

CONTROL: Spray Rogor 0.03 per cent or chloropyriphos 0.02 per cent, Malathion 0.03 per cent

Remarks

According to the Hampson (1976) this species runs close to the *C. hylas* by having following characters:

1. Hind wing cell extremely short.
2. Wings hyaline.

However, it differs from above species by having following characters

1. 3rd and 4th segments are not bright red blackish brown.
2. Collar whitish grey.
3. Penaltimate segment of abdomen with two white spots ventrally.
4. Antenna longer than fore wing.

GENUS *THERETRA*

Differs from Charocampa in having the basal joints of the palpi hollowed out with an orifice towards exterior beset with sensory setae.

Theretra nessus Drury (Figure 87)

Male

Body 55 mm long, 9 mm broad, antenna 18 mm, hind leg 36 mm long, wing expanse 34.5 mm.

Head

Large, 6 mm long, 4 mm broad, green suffused with ferroginous and blackish; eyes rounded black, occular distance 2 mm; antenna 18 mm long, yellow coloured, scape 0.7 mm, pedicel 0.5 m, flagellum 16.8 mm.

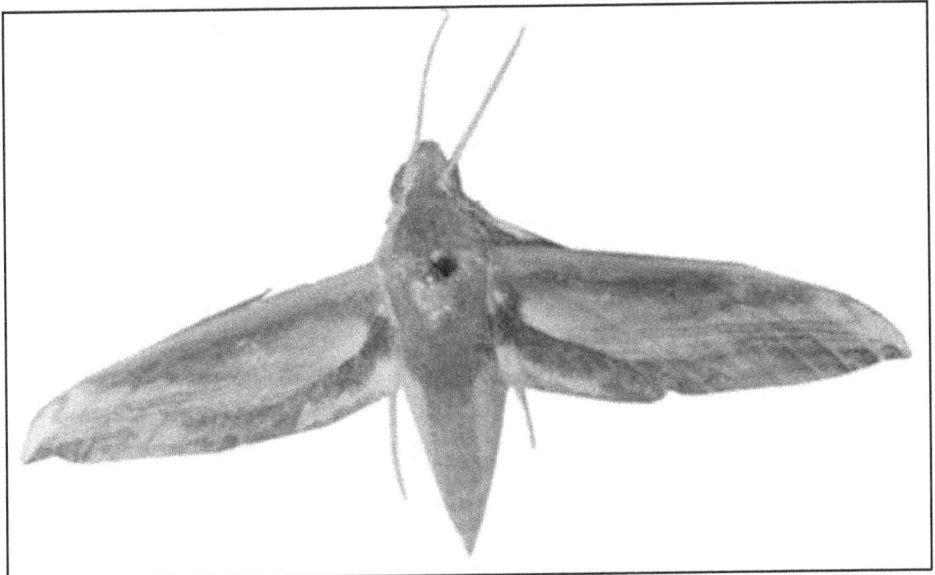

PLATE 26—Figure 87: *Theretra nessus* Drury.

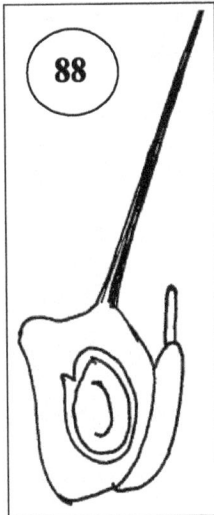

Figure 88: Head and Appendages.

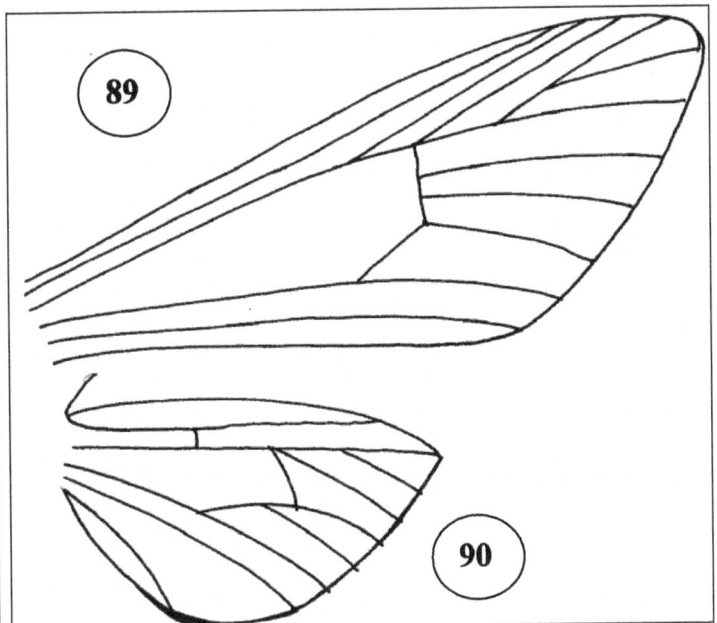

Figure 89: Fore Wing Venation.

Figure 90: Hind Wing Venation.

Thorax

13 mm long, 9 mm broad, blackish with lateral grey strip. **Fore wing:** 50mm long, 18 mm broad, olive brown the base green with a patch of green colour, a black dot at end of cell, a postmedial waved oblique line, met by three straight oblique lines form the apex at inner margin, two submarginal lines, wing expanse 113 mm. **Hind wing :** 25 mm long, 15 mm broad, black brown, the anal angle ochreous which colour extends towards the apex as submarginal band, underside suffused with reddish ochreous. **Hind leg :** 36 mm long, coxa 7 mm, trochanter 1 mm, femur 10 mm, tibia 12 mm, tarsus 6 mm, claws curved and sharp.

Abdomen

36 mm long, 9 mm broad, blackish brown dorsally, yellowish ventrally and golden yellow at sides, a strip down centre of abdomen green suffused.

Female

Similar to male except sexual characters and larger in size.

Host Plants

Citrus plant.

Holotype

Female, India, Maharashtra, Shahuwadi, 9 VIII 2011, V. Y. Kadam, pinned insect in insect box and labeled as above.

Paratype

Female 4, male 2, sex ratio (M:F) 1:2, coll. Maharashtra, Radhanagari, from June to December.

Distribution

Western Ghats of Maharshtra, Kolhapur and Satara, throughout India, Shri Lanka, Myanmar, Hongkong, Borneo, Java, etc.

Control

Spray Rogor 0.03 per cent or chloropyriphos 0.02 per cent, Malathion 0.03 per cent.

Remarks

According to Hampson (1976) this species is *Theretra nessus* Drury.

GENUS : *DELEPHILA*

Delephila ochs. Eur schmett iv, P. 42 (1816)

Type, D livornica, Esp.

Range, Himalayas, China, Hawai, N and S America.

This genus is characterised by

1. The end of each antenna being much thicker, with the hook very short.
2. Outer margin of fore wing as in typical Chaerocampa.

From this genus *Dilephila euphorbia* Linn., *Dilephila galii* and *Dilephila dahlla* species have been described.

Dilephila indica sp.nov. (Figure 91)

Female

Body 38 mm long, 9 mm broad, blackish, antenna 12 mm long, hind leg 24.7 mm long, wing expanse 73 mm.

Head

4 mm long, 3 mm broad, brownish with white striped; eyes oval black and large; frons blackish, occular distance 1.6 mm, collar white striped; antenna 12 mm long, 0.3 mm broad, black, scape 0.6 mm, pedicel 0.4 mm, flagellum 11 mm.

Thorax

11 mm long, 9 mm broad, brown coloured. **Fore wing** : 32 mm long, 11 mm broad blackish, white transverse stripes on fore wing. **Hind wing :** 18 mm long, 10 mm broad, anal margin pinkish in colour, base blackish brown at tip, 4 series of white elongated marginal spots and sub marginal large blackish pathches. **Hind leg :** 24.7mm long, coxa 4mm, trochanter 0.7mm, femur 7 mm, tibia 6 mm, tarsus 7 mm.

Abdomen

23 mm long, 8 mm broad, brownish dorsally and ventrally.

Male

Similar to female except sexual characters.

Host Plants

T. indica

Holotype

Female, India, Maharashtra, Koyanangar, 1 VIII 2010, V. Y. Kadam, pinned insect in insect box, lableled as above.

Paratype

Female 2, male 2, sex ratio (M: F) 1:1, coll. Maharashtra, Pachgani, from June to December, V. Y. Kadam.

Distribution

Western Ghats of Maharashtra, Satara and plain region of Kolhapur.

Control

Spray Rogor 0.03 per cent or chloropyriphos 0.02 per cent or Malathion 0.03 per cent.

Remarks

According to Hampson (1976) this species runs close to *D. livornica* ESP. However, it differs from above species by having following characters.

PLATE 27—Figure 91: *Dilephila indica* sp.nov.

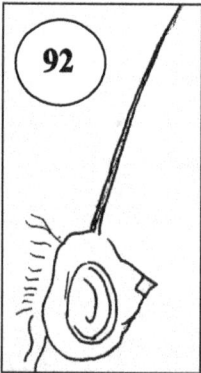

Figure 92: Head and Appendages.

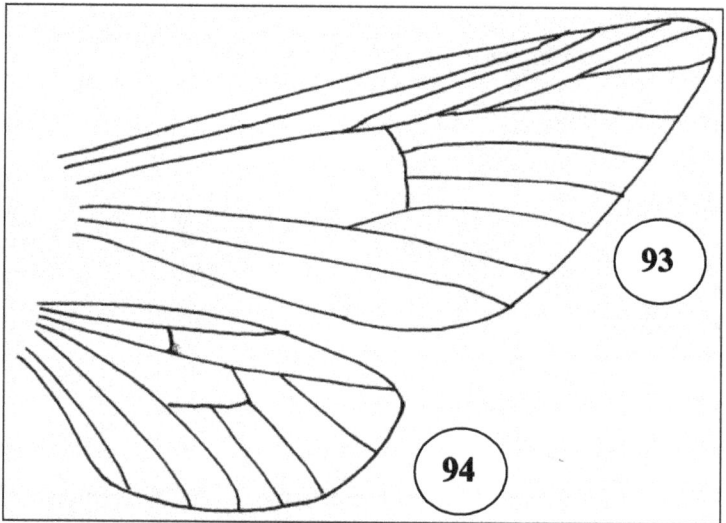

Figure 93: Fore Wing Venation.

Figure 94: Hind Wing Venation.

1. Collar brown with white strip.

2. Abdominal Segment 4th and 5th with two white strips in pair.

3. Anal area of hind wing pinkish.

4. White elongated broken stripes on fore wing.

FAMILY : NOCTUIDAE

Noctuidae is large family of moths consisting of closely allied subfamilies and genara. Almost all are of completely nocturnal habits except the Deltoidinae.

The family is characterised by fore wing with vein 1 a straight and not anstomosing with 1b, 1c absent, 2 from middle of cell, 3, 4, 5 from close to lower angle. 6 from upper angle, 8 given off from 7 and anastomosing with 9, which is given of from 10 to form an areole, 11 from cell, 12 long. Hind wings with 1a and 1b present, 1c absent, 2 from middle of cell, 3 and 4 from lower angle and 5 from near lower angle or middle of discocellulars, rarely absent. Frenulum and proboscis always present.

According to Hampson (1976) larvae usually nacked or slightly clothed with hair, rarely with thick tuft or spatulate filaments. The lowest froms, the Sarrothripinae, Stictopterinae, Trifinae, Palindiinae, Eucteliinae and almost all Genopterinae have the four pairs of abdominal prolegs fully developed and are not semilooping, whilst in the Deltoidinae and Acontiinae the first or first two pairs are sometimes absolescent or absent, in the Quadifinae and Facilinae, which are semiloopers, the first or first two pairs are almost always rudimentary or absent. The larvae usually pupute in the earth without a cocoon, a cemented chamber, being often formed more rarely a cocoon is made amongst leaves, on bark, or on the surface of ground.

Hampson (1976) visualized following 10 sub families under the family Noctuidae

1. Trifinae

2. Acontiniae

3. Gonopterinae

4. Euteliinae

5. Sarrothripinae

6. Stictopterinae

7. Palindiinae

8. Quadrifiinae

9. Focillinae

10. Deltoidinae

GENUS *HAMODES*

Hamodes, Guen. Noct iii, P. 202 (1852)

Armana, Swinh. Trans. Ent. Soc., 1890 P.250.

Kalmina, Swinch. Trans. Ent. Soc., 1891. P.480.

Type, *H. propitia*, Guen, from New Ireland.

Range : Throughout India and Myanmar, Andamans, New Ireland.

This genus is characterized by:

1. Palpi reaching vertex of head, the 3rd joint minute.
2. Antennae minutely fasciculated in males.
3. Head and thorax smoothly scaled.
4. Tibiae not hairy and without spines.
5. Fore wings usually highly arched towards apex.
6. Hind wing with anal angle truncate. Vein 5 from lower angle of cell. From India only four species of this genus have been studied.

KEY TO THE SPECIES OF THE GENUS *HAMODES*.

1. Male with no tuff on underside of fore wing. Hind wing with costa not arched   1

    Fore wing of male with large tuft of hair on underside below end of cell.

    Hind wing with costa immensely arched   2,3

2. Fore wing of outer margin oblique

    The costa highly arched, collar black

    Fore wing irrorated with fuscus   *aurantiaca*

    Fore wing with outer margin errect, collar red.

    Fore wing with black spots in cell   *unilines*

    Fore wing with black kidney shaped spot and with oval elongated spots larger towards margin side, smaller towards inner side, pale yellow in appearance. Abdomen whitish   *shivajinensis*

    With kidney shaped black spot on fore wing. 3 blacks spots, larger towards inner side, smaller toward apical side brownish in appearance, abdomen orange   *indica*

    Bright orange fore wing with no black spot on costa above the reniform   *ochracea*

3. Tibiae more hairy fore wing lost slightly arched, with minute black spots in cell   *nigriricta*

Hamodes shivajiensis sp.nov. (Figure 95)

Female

Body 29 mm long, 7 mm broad, light grey dorsally and yellowish ventrally; antenna simple, 23 mm long, hind leg 23 mm long.

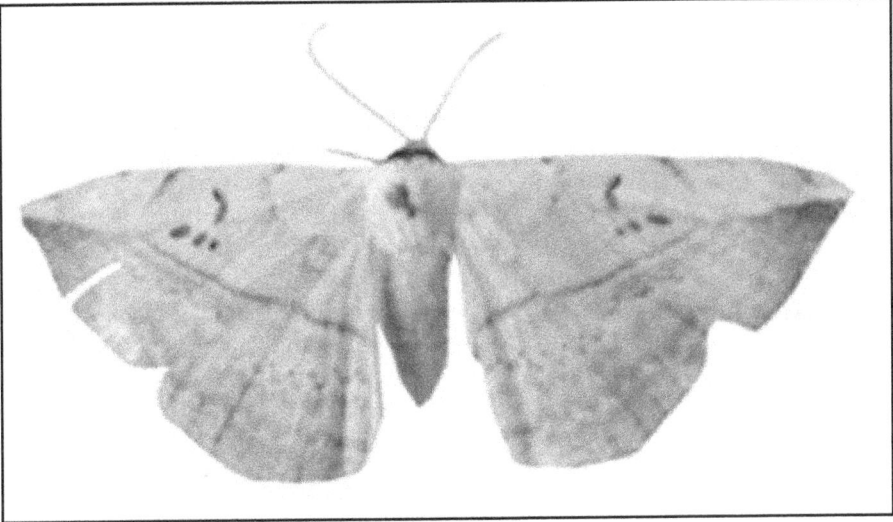

PLATE 28—Figure 95: *Hamodes shivajiensis* **sp.nov.**

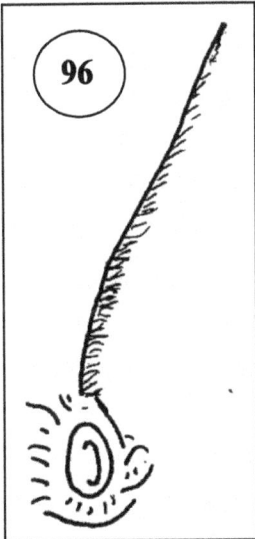

Figure 96: Head and Appendages.

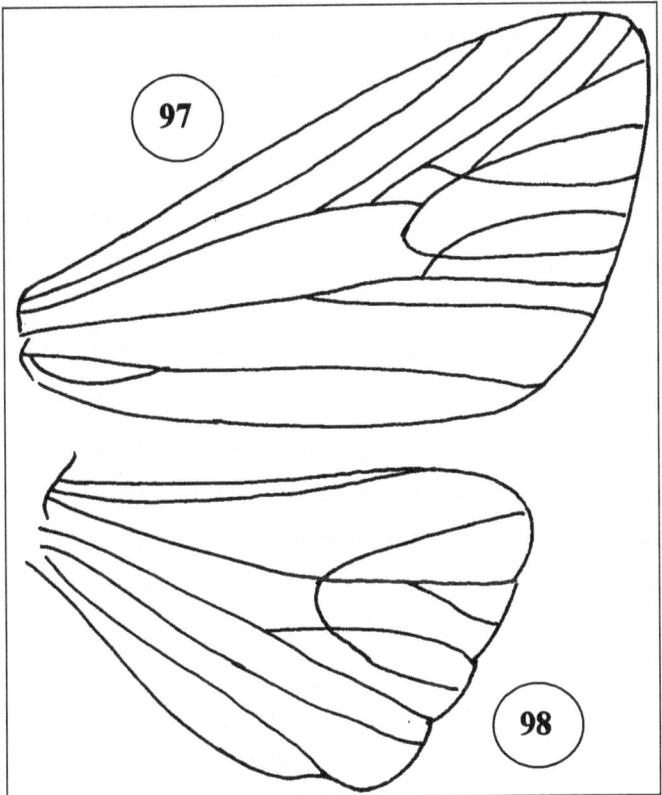

Figure 97: Fore Wing Venation.
Figure 98: Hind Wing Venation.

Head

2 mm long, 1.3 mm broad, light grey, dorsally and orange ventrally; eyes large, occular distance 0.8 mm, collar brown; antenna 23 mm long, 0.2 mm broad, pointed at tip, brown coloured, scape 0.7 mm, pedicel 0.4 mm flagellum 21.9 mm.

Thorax

11 mm long, 7 mm broad, light grey dorsally and orange ventrally. **Fore wing:** 40 mm long, 20 mm broad, costa highly arched, light grey dorsally with 1 large, black kidney shaped spot, 3 spots at mid region larger from outer and smaller from inner region, serrate margin, ventrally dark yellow with three zig-zag black markings. **Hind wings :** 25 mm long, 22 mm broad, grey dorsally and yellow ventally. **Hind leg :** 23.6 mm long, coxa 6 mm, trochanter 0.6 mm, femur 9 mm, tibia 4mm, without spurs, tarsus 3mm.

Abdomen

16 mm long, 7 mm broad, whitish pale yellow dorsally and dark yellow ventrally, two black spots on last abdominal segment, ventally one black spot.

Male

Antennae minutely fasciculated.

Host Plant

Ficus glomerata.

Holotype

Female, India, Maharashtra, Koyananagar, coll. 21 VII 2011, V. Y. Kadam.

Paratype

Female 2, male 1, sex ratio (M:F) 1: 2, Radhanagari, coll. July to November, T. V. Sathe.

Distribution

Western Ghats of Maharashtra, Satara and Kolhapur.

Control

Spray Rogor 0.03 per cent or Malathion 0.03 per cent or Carbaryl 0.03 per cent.

Remarks

According to Hampson (1976) this species runs close to *Hamodes qurantiaca* by having dark kidney shaped spot on the wing, costa highly arched.

However, it differs form above species by having following characters

1. Three black spots on the fore wing, larger at outer side and smaller at inner side.
2. Collar brownish black.
3. Abdomenal last segment with two black spots.

Hamodes indica sp.nov. (Figure 99)

Female

Body 29 mm long, 5 mm broad, light grey dorsally and yellowish ventrally; antenna filiform, 20 mm long, hind leg 23 mm long, wing expanse 79 mm.

Head

3mm long, 2 mm broad; eyes large round, occular distance. 0.8 mm, proboscis long coiled; antenna 20 mm long, scape 0.6 mm, pedicel 0.4 mm, flagellum 19 mm.

Thorax

Collar blakish brown, 7 mm long, 5 mm broad, dorsally light grey, ventraly yellowish. **Fore wing :** 37 mm long, 22 mm broad, 3 black spots and without Kidney shaped spot at mid region, serrate margin 4 zig-zag lines distally, **Hind wing :** 25mm long, 21 mm broad. **Hind leg :** 23 mm long, coxa 5 mm, trochater 1mm, femur 9 mm, tibia 4 mm, tarsus 4 mm, claws curved and reduced.

Abdomen

20 mm long, 4 mm broad, light grey dorsally and yellowish ventrally tip yellow in colour. Last four segments light yellow, rest segments whitish.

Male

Similar to female except sexual characters.

Host Plants

T. indica

Holotype

Female, India, Maharashtra, Panchgani, coll. 15 VII 2010, T.V. Sathe.

Paratype

Female 2, male 2, India, Maharashtra, Shahuwadi, coll. June to December, V. Y. Kadam.

Distribution

Western Ghats of Maharashtra, Kolhapur, Satara.

Control

Spray Rogor 0.03 per cent or Malathion 0.03 per cent or Carbaryl 0.03 per cent.

Remarks

According to Hampson (1976) this species runs close to *Hamodes nigriricta* by having following characters:

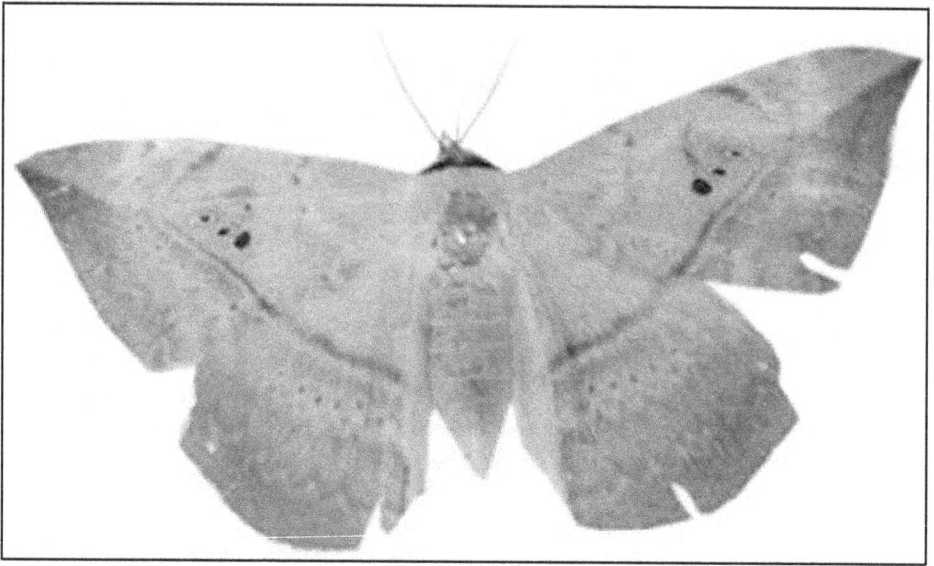

PLATE 29—Figure 99: *Hamodes indica* sp.nov.

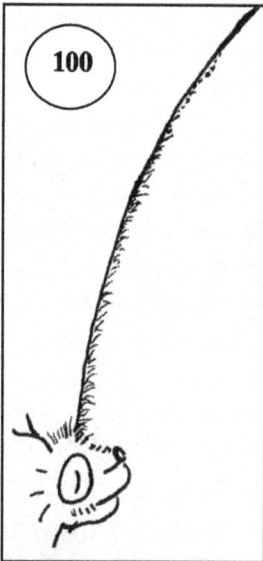

Figure 100: Head and Appendages.

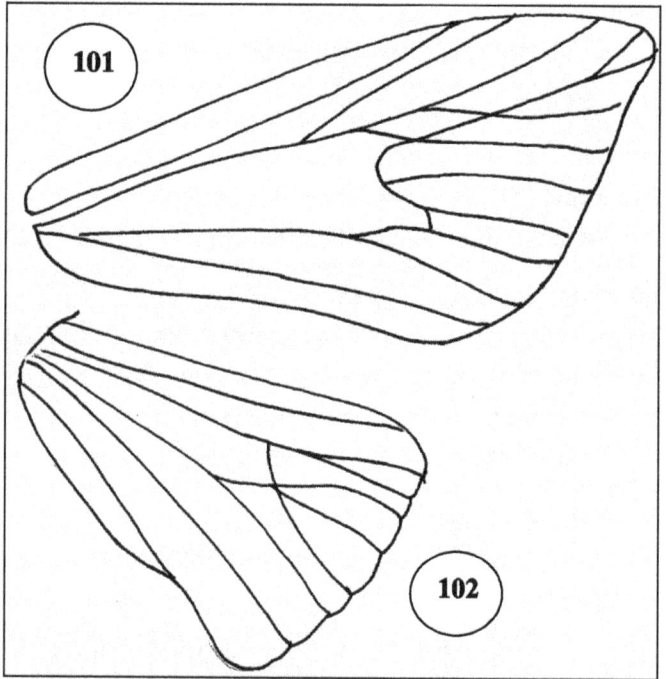

Figure 101: Fore Wing Venation.
Figure 102: Hind Wing Venation.

1. Tibiae more hairy.
2. Male with no tuft on under surface of fore wing. Hind wing with costa not arched. However, it differs from above species by having following characters
   1. Fore wing without black kidney shaped spot.
   2. 3 blacks spot on fore wing larger towards inner side.
   3. Body brownish in appearance.
   4. Abdominal last segment with two black spots and segment yellow coloured.

GENUS *ELIGMA*

Eligma, Hubn. verz. P. 164 (1818)

Panglima, Moore, Lep. E.I. Co. P. 297 (1858)

Surina, Wlk. Proc. Nat. Hist. Soc. Glasgow, pp. 333(1969)

Type, *E. narcissus.* Cram.

Range : Africa, China, throughout India, Shrilanka, Penang, Java.

The genus *Eligma* is charactered by

1. Palpi upturned, the second joint reaching above vertex of head, the third very long and slender and some what spatulate at tip.
2. Antennae with cilia minute.
3. Tibia thickly scaled.
4. Fore wing narrow, the costa much arched at base, the outer margin nearly erect to vein 3, then obliqua to outer angle, vein 3,4,5 from close to lower angle of cell, 6 from just below upper angle, 7 and 10 from a long areole formed by the anastomosis of 8 and 9.
5. Hind wing with veins 3,4,5 from close to lower angle of cell, 6 and 7 from upper angle.

From India only one species have been reported under this genus.

Eligma narcissus Crammer (Figure 103)   .

Eligma narcissus Cram : Pap.Exot. 1, Pl. 73, Figure E. F. C and S No. 559.

Female

Body 21 mm long, 4 mm broad, wing expanse 66 mm, antenna simple, 14 mm long.

Head

2.56 mm long, 2 mm broad, grey brown; eyes large, black oval; frons grey, palpi short, occular distance 0.7 mm; antenna 14 mm long, black, with cilia minute, scape 0.7 mm, pedicel 0.5 mm, flagellum 12.8 mm long.

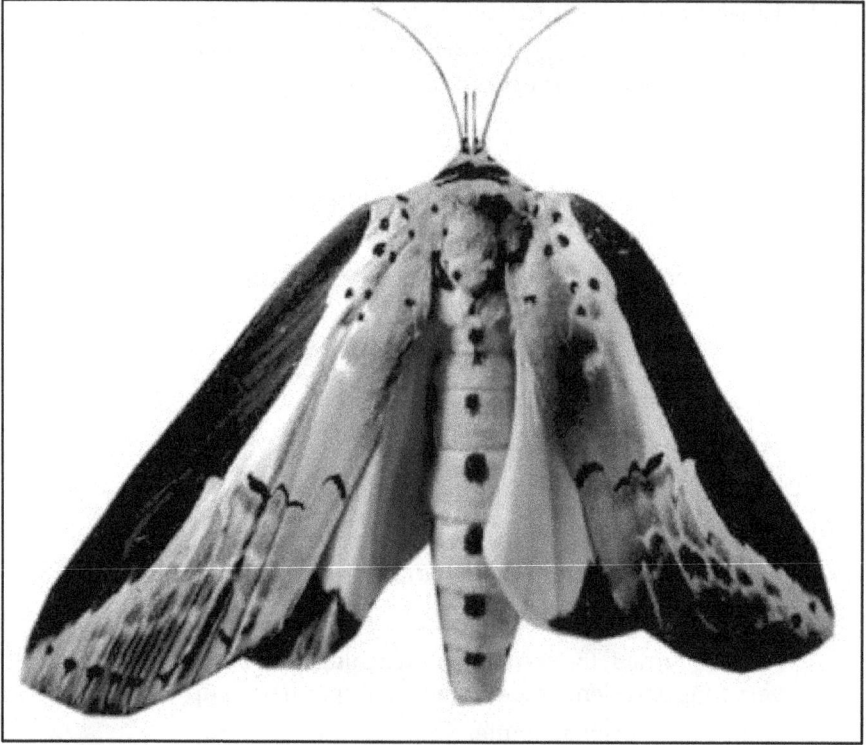

PLATE 30—Figure 103: *Eligma narcissus* **Crammer.**

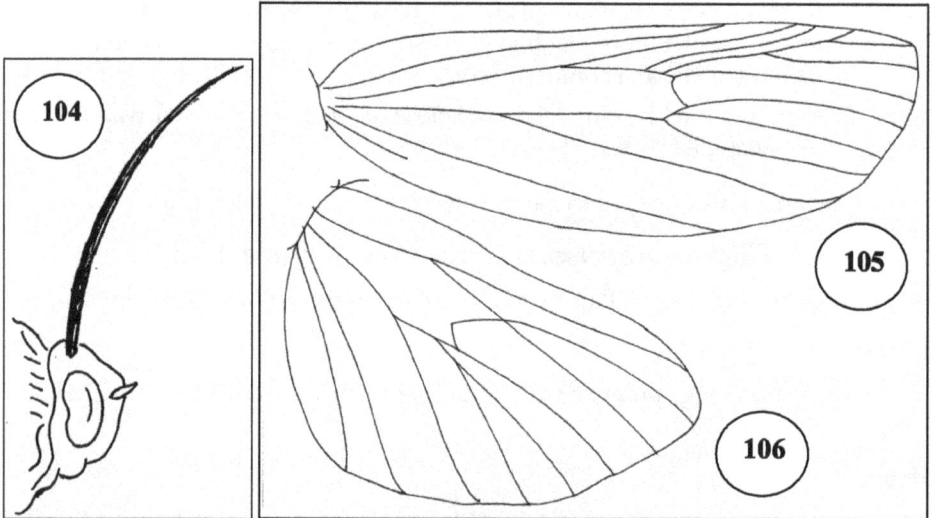

Figure 104: Head and Appendages.

Figure 105: Fore Wing Venation.
Figure 106: Hind Wing Venation.

Thorax

5 mm long, 4 mm broad, grey brown, black spotted and yellow ventrally. **Fore wing :** 31 mm long, 12 mm broad, grey brown, the costal area with an olive tinge, an irregular and diffused curved white, fascia from base to apex. Some basal and subbasal black spots, a wave black line from lower angle of cell to inner margin, submarginal series of black spots, towards outer angle conjoined into streaks. **Hind wing :** 23 mm long, 14 mm broad, bright yellow dorsally and ventrally. The apical area black suffused or streaked with dark blue, cilia white at tips.

Hind leg : 17.1 mm long, coxa 5 mm, trochanter 0.7 mm, femur 6 mm, tibia 3 mm, tarsus 2.4 mm, black spots on the tibia, the tarsi grey brown.

Abdomen

14.5 mm long, 3mm broad, dorsally bright yellow with dorsal and lateral series of black spots.

Male

Similar to female except sexual characters and smaller in size.

Host Plants

Ailanthus triphysa, Ailanthus excelsa.

Holotype

Female, India, Maharashtra, Radhanagari, coll. 9 VIII 2010, V. Y. Kadam, pinned insect in insect box and labeled as above.

Paratype

Female 10, male 4 (M:F) 1:2.5, coll. from July to December, V. Y. Kadam, Radhanagari.

Distribution

All over India, China, Myanmar, Andamans, Malaysisa, India western Ghats of Maharashtra, Kolhapur, Satara.

CONTROL: Spray Rogor 0.03 per cent/Malathion 0.03 per cent/Azadirachtin 0.03 per cent.

Remarks

According to Hampson (1976) this species is *Eligma narcissus* Crammer. However, the species is poorly described previously. Hence, redescribed in the text.

GENUS *HELIOTHIS*

Heliothis ochs Eur. Schmett iv P. 91 (1816)

Heliocheilus, Grote, Proc. Ent. Soc. Phil. iv. 1865, P. 328

Type, *H. dipsacea.* Linn., from Europe

Range : Universally distributed.

The genus *Heliothis* is featured by

1. Eyes naked and without lashes.
2. Proboscis fully developed.
3. Palpi porrect, the second joint evenly clothed with long hair, the third short and depressed, a short forontal tuft.
4. Thorax and abdomen without tufts.
5. Fore tibia with a pair of slender terminal spines.
6. Mid and hind tibiae spined.
7. Fore wing with vein 8 and 9 sometimes given off from the end of areole. From India, 5 species have been reported and described under the genus *Heliothis.*

Heliothis nocturni sp.nov. (Figure 107)

Female

Body 18 mm long, 4 mm broad, reddish brown, antenna 13 mm long, hind leg 16.3mm long, wing expanse 44 mm.

Head

Short, 2 mm long, 1.5 broad, brown coloured; eyes large oval, occular distance 9.8 mm, collar blackish brown; frons brownish; antenna 13 mm long, 0.2 mm broad, simple filiform, scape 0.6 mm, pediciel 0.4 mm, flagellum 12 mm, black coloured.

Thorax

6 mm long, 4 mm broad, yellowish. **Fore wing :** 20 mm long, 11 mm broad, dark brown coloured, serrate margin, male without hyaline patches on fore wing, 3 transverse black bands, out of which marginal more prominent. **Hind wing :** 16 mm long, 10 mm broad blackish, without hyaline patches. A dark large blackish brown band at marginal level and transverse yellow band, rest of the wing is light brown. **Hind leg :** 16.3 mm long, coxa 4 mm, trochanter 0.3 mm, femur 5 mm, tibia 3 mm, tarsus 4 mm.

Abdomen

10 mm long, 3 mm broad, brownish coloured, transverse white bands, ventrally yellowish. Terminal segment organge yellow without black spot.

Male

Simlar to female except sexual characters.

Host Plants

Pulses, *Fagiolus* sp.

Holotype

Female, India, Maharashtra, Shahuwadi, coll. 3 VII 2009, V. Y. Kadam, pinned insect in insect box and labeled as above.

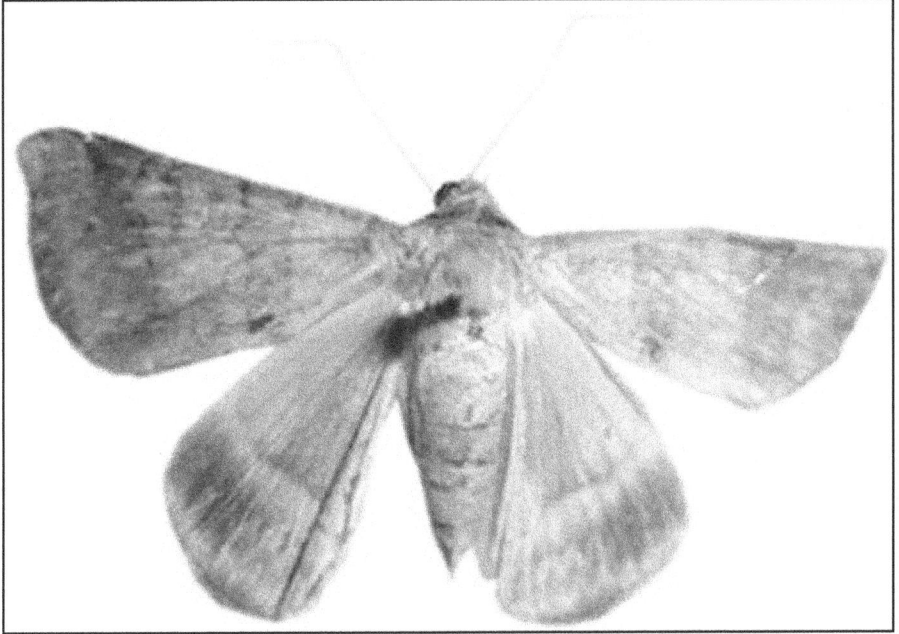

PLATE 31—Figure 107: *Heliothis nocturni* **sp.nov.**

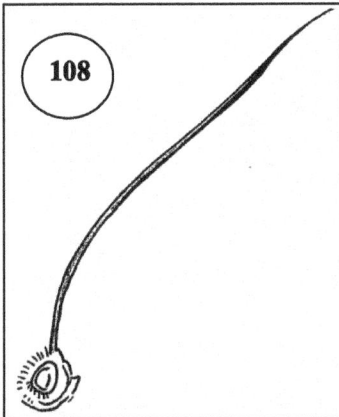

Figure 108: Head and Appendages.

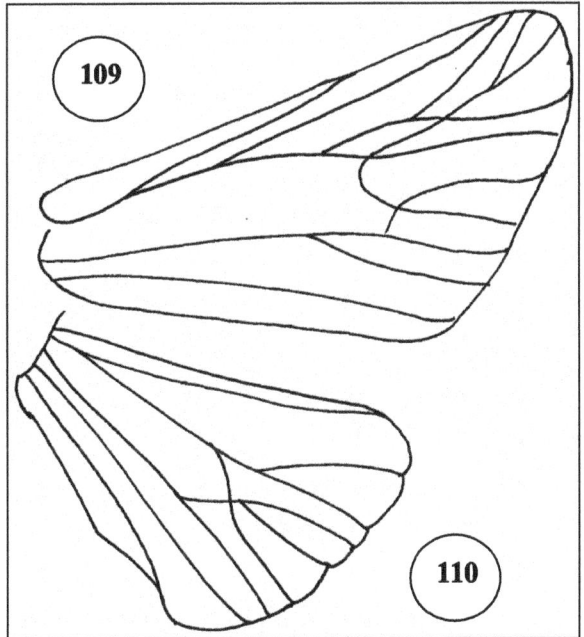

Figure 109: Fore Wing Venation.

Figure 110: Hind Wing Venation.

Paratype

Female 1, male 1, sex ratio (M:F) 1:1, coll. from July to April, T. V. Sathe.

Distribution

Western Ghats of Maharashtra, Kolhapur, Satara.

Control

Spray Rogor 0.03 per cent/Malathion 0.03 per cent/Azadirachtin 0.03 per cent.

Remarks

According to Hampson (1976) this species runs close to *Heliothis scutosa* Schitt. by having following characters:

1. Body reddish brown.
2. Male without hyaline wings.

However, it differs from above species by having following characters

1. Collar black brown
2. Fore wing with 3 transverse blackish bands, out of which marginal is more prominent.
3. Hind wing with a dark large blackish brown band at submarginal level and one yellow band, rest of the wing with dark brown.
4. Abdominal last segment orange yellow more prominent ventrally.

Heliothis armigera Hubn. (Figure 111)

Heliothis armigera Hubn. Samml. Eur. Schmett. Noct. Ii Pl. 79.

Figure 320, C.and S. no. 1730, Hampson III Pl. 176. Figure 22.

Helothis pulverosa Wlk cat. Xi, P. 688.

Heliothis conferta Wlk Cat. XI P. 690.

Heliothis succinea Moore P.Z. S. 1881, P. 362, C. and S no. 1738.

Heliothis rubrescens C and S. no. 1736 (nec wlk.)

Helicoverpa armigera Hubn.

Female

Body 20 mm long, 3 mm broad, ocheous with pale brown, antenna 11 mm long, hind leg 12.9 mm, wing expanse 43 mm.

Head

3 mm long, 2 mm broad; eyes round black coloured mustard like occular distance 1.5 mm; frons white, collar, brownish; antenna 11 mm long, scape 0.7, pedicel 0.4 mm, flagellum 9.9 mm.

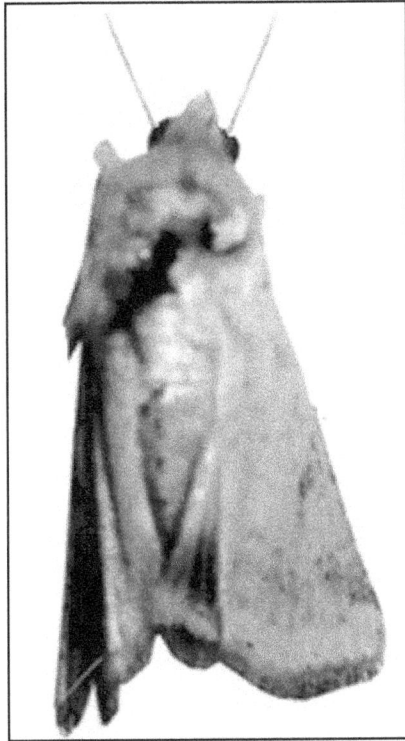

PLATE 32—Figure 111: *Heliothis armigera* **Hubn.**

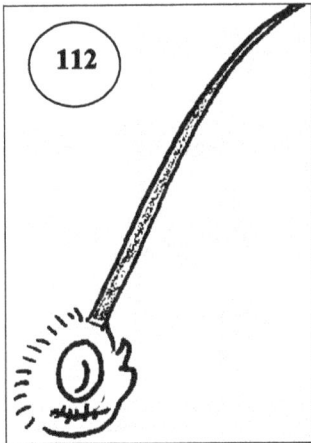

Figure 112: Head and Appendages.

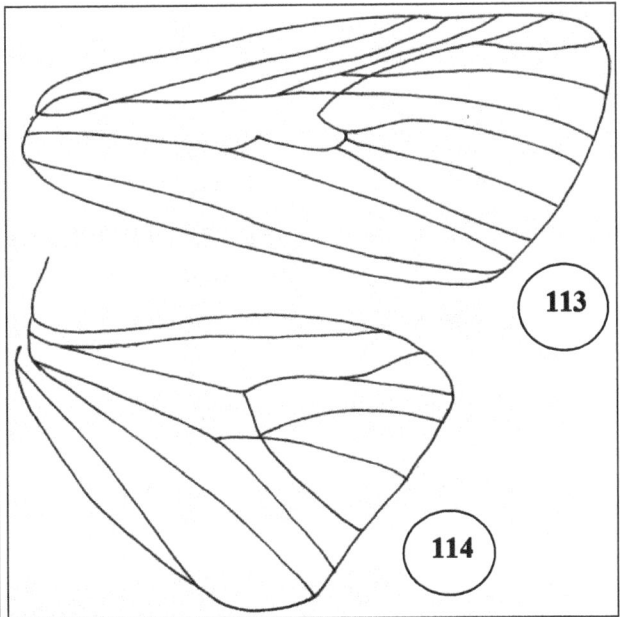

Figure 113: Fore Wing Venation.

Figure 114: Hind Wing Venation.

Thorax

5 mm long, 3 mm broad, straw coloured, wing expanse 43 mm. **Fore wing :** 19 mm long, 9 mm broad with double indistinct waved antimedial lines, small circular spot at middle of the fore wing, post medial and submarginal waved lines, marginal series of dark speakes. **Hind wing :** 11 mm long, 7 mm broad, white, blackish outer border with marginal dark large patches. The apices of both wings and outer area of fore wing pinkish. **Hind leg** : 12.9 mm long, coxa 2.0 mm, trochanter 0.4 mm, femur 4 mm, tibia 3 mm, tarsus 3.5 mm.

Abdomen

12 mm long, 3mm broad, straw coloured, without black spots.

Host Plants

Gram, *Cecer arientum, Cajanous cajan* Mill. *Sorghum valgare.*

Holotype

Female, India, Maharashtra, Gaganbawada, Coll. 17-XI 2010, T. V. Sathe.

Paratype

Female 2, male 1, India, Maharashtra, Shahuwadi, coll. from June to December, V. Y. Kadam, coll. throughout year.

Distribution

India, Cosmopolitan, western Ghats of Maharashtra, Kolhapur, Satara.

Control

Spray Rogor 0.03 per cent/Malathion 0.03 per cent/Azadirachtin 0.03 per cent.

Remarks

According to Hampson (1976) this species is *Heliothis* (= *Helicoverpa*) *armigera* Hubn. now synonymised as *Helicoverpa armigera* Hubn.

GENUS *PRODENIA*

Prodenia, Guen. Noct. I, p. 159 (1852)

Type, *P. littoralis,* Buisd.

Range : Nearctic region, mediterraneam subregion and tropical and subtropical zones. This genus is characterised by having tufts on metathorax, scaling much smoother, legs less hairy, abdominal tufts slight, antennae ciliated in male.

Prodenia (Spodoptera) littoralis Fab. (Figure 115)

Female

Body 15 mm long, 4 mm broad, pale ochreous much suffused with dark brown, antenna 10 mm, hind leg 11.8 mm, wing expanse 40 mm.

PLATE 33—Figure 115: *Prodenia (Spodoptera) littoralis* **Fab.**

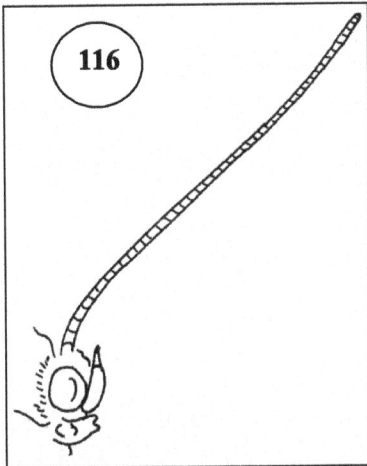

Figure 116: Head and Appendages.

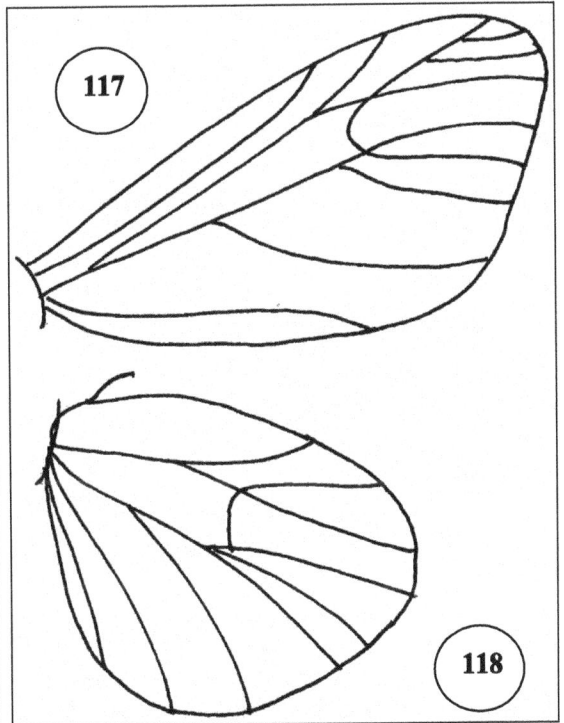

Figure 117: Fore Wing Venation.

Figure 118: Hind Wing Venation.

Head

2 mm long, 1.5 mm broad, ochreous; eyes large black rounded, occular distance 1.1 mm; frons black, collar light brownish; antenna 10 mm long, scape 0.6 mm, pedicel 0.5 mm, flagellum 9.9 mm.

Thorax

4 mm long, 3 mm broad, ochreous, wing expanse 40 mm. **Fore wing :** 18mm long, 7 mm broad, brown blackish spot, with zig-zag golden marking with ochreous, streaks at base, an angled and oblique subbasal line reniform an oblique "arrow head" mark, indistinct lunalate post medial line excurved beyond cell. **Hind wing :** 13 mm long, 9 mm broad, whitish, semihyaline, with dark marginal line. **Hind leg:** 11.8 mm long, coxa 2 mm, trochanter 0.3 mm, femur 5mm, tibia 2 mm, tarsus 2.5 mm.

Abdomen

9 mm long, 3 mm broad, blackish.

Male

Similar to female except sexual characters.

Host Plants

Polyphagous, Tobacco, Groundnut, Sal, Teak, Castor, etc.

Holotype

Female, India, Maharashtra, Kolhapur, Radhanagari, Coll. 18-VIII 2011, T. V. Sathe, specimen pinned and labeled as above.

Paratype

2 females, 2 males, coll. India Maharashtra, Kolhapur, Gaganbawada, from July to May, V. Y. Kadam.

Distribution

Western Ghats of Maharashtra, Kolhapur, Satara. Distributed throughout India.

Control

Spray Rogor 0.03 per cent/Malathion 0.03 per cent/Azadirachtin 0.03 per cent.

Remarks

Accroding to the Hampson (1976)) this species *is Prodenia littoralis* Buisd. However, recently it has been kept under the genus *Spodoptera* as *Spodoptera litura* Fab.

FAMILY – HYPSIDAE

The family Hypsidae is closely related to tiger moths, Arctiidae but can be separated by presence of an areole on fore wing at the origin. The family Hypsidae is also characterised by:

1. Presence of proboscis, palpi smoothly scaled, the 3rd joint long and naked.
2. Legs smooth and tibia with one pair of spurs.
3. Hind tibia with two pairs.
4. Presence of frenulum.
5. Fore wing with vein l a separate from 1 b, 1 c absent, 5 from near the lower angle of cell.
6. A ridge of membrane below the costa of hind wing.
7. Hind wing with vein l a and 1 b present, lc absent; 5 from near lower angle of cell; 8 free from the base and connected by a bar with 7 at middle of cell.

   20 species of the genus Hypsa have been described from India by Hampson (1976).

Key to the Genera

1. Fore wing with no areole, vein 6 and 7 stalked   *Euplocia*
2. Palpi with the second joint reaching above vertex of head   *Peridrome*
3. Hind wing with vein 6 and 7 from angle of cell   *Hypsa*
4. Fore wing with vein seven from and areole.
5. Palpi with second joint reaching vertex of head
6. Hind wing with vein 6 and 7 stalked   *Digama*

KEY TO THE LOCAL SPECIES OF THE GENUS *HYPSA*

1. Antennae of male with fasciculated cilia short

   Palpi with 3 rd joint of moderate length   1

   Palpi with 3 rd joint very long   2

   Palpi with 3 rd joint very short   3

2. Palpi black spots on 1 and 2 joints. Hind wing orange yellow, a black spots on the end of cell   *alcipron*

   A large black spots at the centre of hind wing and small 3 black spots on costal margin and small black spots forming S shaped series of at margin   *marathi sp.nov.*

3. Hind wing bright orange yellow with apical black band. Two black spots on costa   *ficus*

4. Head black and grey, thorax orange, collar banded with black   *marmorea*

Hypsa producta Moore (Figure 119)

Female

Body 22 mm long, 4 mm broad, wing expanse 54 mm, antenna 14 mm long.

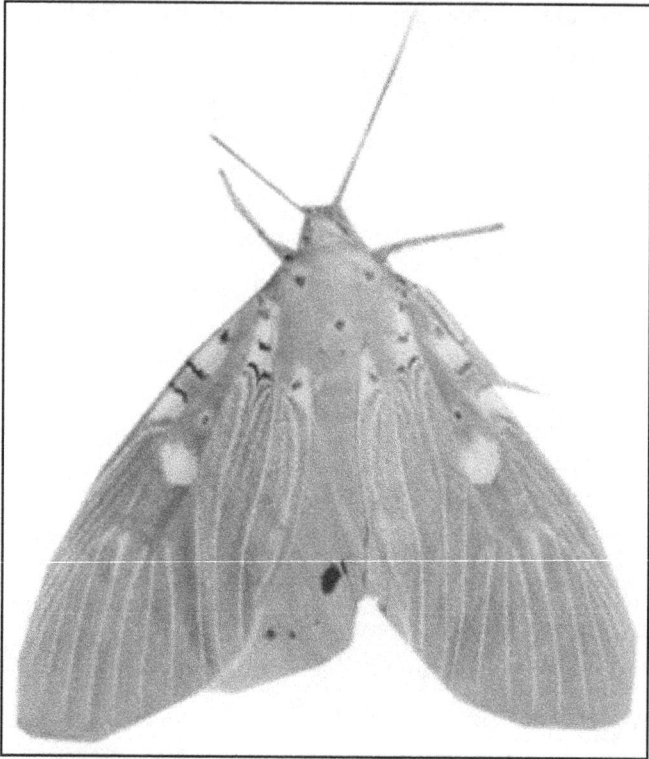

PLATE 34—Figure 119: *Hypsa producta* Moore.

Figure 120: Head and Appendages.

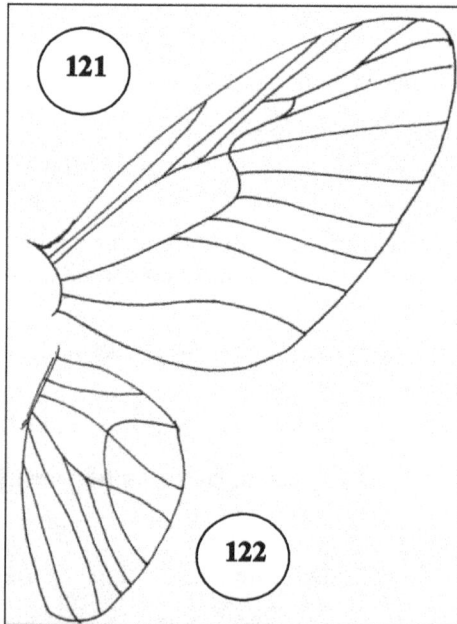

Figure 121: Fore Wing Venation.
Figure 122: Hind Wing Venation.

Head

2 mm long, 1.6 mm broad; eyes small, proboscis coiled; antenna 14 mm long, blackish, scape 0.6 mm, pedicel 3 mm, flagellum 13 mm long.

Thorax

5 mm long, 4 mm broad, smooth, bufi, yellow on dorsal side. **Fore wing :** 25 mm long, 16 mm broad, yellow with 6 white spots at base, brownish distally. **Hind wing :** 21 mm long, 12 mm borad, yellowish with 3 black spots. **Hind leg :** 16 mm long, coxa 5 mm, trochanter 0.4 mm, femur 6 mm, tibia 2 mm, tarsus 2.6 mm, claws reduced.

Abdomen

15 mm long, 4 mm broad.

Male

Antennae with fasciculated large cilia.

Host Plant

F. glomerata.

Holotype

Female, India, Maharashtra, Satara, Koyanangar, 5 VIII 2010, coll. V. Y. Kadam.

Paratype

12 female, 6 male, coll. June to October, Koyananagar, V. Y. Kadam.

Distribution

Through out India.

Control

Spray Rogor 0.03 per cent or Carbaryl 0.15 per cent.

Remarks

According to Hampson (1976) this species is *Hypsa producta.*

Hypsa marathi sp.nov. (Figure 123)

Female

Body 28 mm long, 4 mm broad, yellowish, wing expanse 64 mm, antenna 12 mm long.

Head

2 mm long, 1.4 mm broad; eyes large round; frons yellow, occular distance 1. 2 mm; proboscis coiled, pulpi third joint very long, palpi with black spot on 2nd and 3rd joint; antenna 12 mm long, 0.4 mm broad, simple, brownish, scape 0.6 mm pedicel 0.4 mm, flagellum 11 mm long.

PLATE 35—Figure 123: *Hypsa marathi* sp.nov.

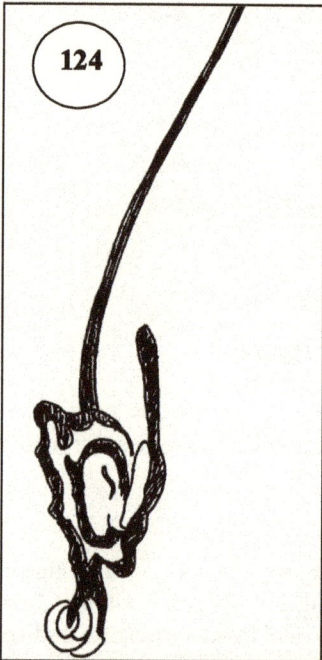

Figure 124: Head and Appendages.

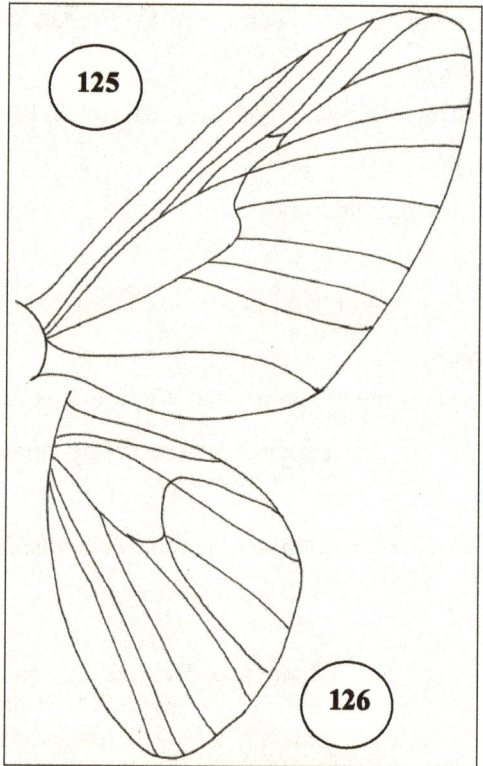

Figure 125: Fore Wing Venation.
Figure 126: Hind Wing Venation.

Thorax

6 mm long, 4 mm broad, smooth yellow buff on dorsal side, collar yelllow. **Fore wing:** 30 mm long, 12 mm broad, brownish fuscus, basal orange patch with two subbasal black spots and series of three on its outer edge. The vein streaked with white, a white spot at lower angle of cell. **Hind wing** : 20 mm long, 17 mm broad, orange yellow black spot at end of cell, one beyond, one below with vein 2 and submarginal irregular series which sometimes became a nearly complete marginal band, the veins crossing it yellow. **Hind leg :** 15.1 mm long, brownish, coxa 5 mm, trochanter 0.4 mm, femur 6 mm, tibia 2 mm, tarsus 1.7 mm, claws reduced and curved.

Abdomen:

20 mm long, 4 mm broad, yellowish with black spots.

Male

Similar to female except sexual characters, antennae with fasciculated cilia short.

Host Plant

F. glomerata.

Holotype

Female, India, Maharshtra, Satara, Panchgani, 2 VIII 2011, V. Y.Kadam, pinned insect in insect box and labeled as above.

Paratype

6 females, 2 males, sex ratio (M.F.) 1:3, coll. Maharashtra, Satara, Panchgani, V. Y. Kadam, from June to September.

Distribution

Throughout India, Sri lanka and Myanmar.

Control

Spray Rogor 0.03 per cent or Carbaryl 0.15 per cent.

Remarks

According to Hampson (1976) this species runs close to *Hypsa alcipron* Cramer by having following characters:

1. Male antenna with fasciculated cilia short and
2. 3rd joint of palpi very long.

However, it differs from the above species by having following characters

1. A large black spot at the centre of hind wing.
2. Hind wing with 3 black spots on costal margin and small black spots forming 'S' shaped series at marginal portion.

3. On abdomen transverse black strips present.

4. Fore wing twice the length of antena.

FAMILY: PYRALIDAE

The family Pyralidae is characterised by

1. Proboscis and maxillary palpi usually developed, fore wing with vein 1a usually free, sometimes forming a fork with 1b; 1 c absent, S from near lower angle of cell; 8, 9 almost always stalked.

2. Hind wing with vein 5 usually from near lower angle of cell. S. approximated to 7 or anastomosing with it beyond the end of cell, 1a, b, c present, frenulam developed.

3. Moths usually slender built with long, thin legs.

4. Larvae elongate, with 5 pairs of prolegs.

5. Pupa with segments 9-11, some times also 8 and 12 moveable, not protruding from cocoon on emergence.

KEY TO SUBFAMILY

Hind wing with the median nervure pectinated on upperside.

1. Fore wing with vein 7 present and Maxillary palpi not triangularly scale.   Galleriinae

2. Maxillery paopli triangularly scaled   Crambinae

3. Fore wing with vein 7 absent and proboscis absent   Anerastriine

4. Fore wing with vein7 absent and proboscis present   Phycitinae

5. Hind wing with median nervar not pectinated on upper side and proboscis absent   Schaenobiinae

6. Fore wing with vein 7 stocked with 8,9 and fore wing with tufts of raised scales in the cell   Epipaschiinae

7. Fore wing with no tufts of raisedscales in the cell and maxillary palpi absent   Chrysauginae

8. Hind wing with vein 8 anstomosing with 7 and maxillary palp present   Endotrichinae

9. Hind wing with vein 8 free   Pyralinae

10. Fore wing with vein 7 form the cell and vein 10 stalked with 8, 9   Hydrocampinae

11. Fore wing with tuft of raised scales in te cell   Scoparinae

12. Fore wing with no raised tufts of scales in the cell   Pyraustinae

KEY TO THE GENERA OF PYRAUSTINAE

1. Fore wing- with veins 7, 8 stalked. Hind wing1 with three veins arising from median nervure. Hind wing with the cell open, vein 7 anastomosing with 8 to about two-thirds of wing Fore wing with vein 10 stalked with 7, 8, 9   *Trachylepidia*

2. Fore wing with vein 10 from cell. Fore wing with vein 7 arising from 8 after 9 2   *Thagora*

3. Fore wing with vein 7 arising from 8 before 9 3   *Muctalla*

4. Hind wing with the cell closed; vein 7 anastomosing with 8 almost to apex 4   *Achroia*

5. Hind wing with four veins arising from median nervure. Hind wing with the cell open   *Lamoria*

6. Hind wing with the cell closed. Palpi clothed with very long hair; fore wing with the outer margin evenly curved   *Acara*

7. Palpi smoothly scaled; fore wing with the outer margin angled *Galleria*

8. Fore wing with vein 7 arising from cell; hind wing with four veins from median nervure, 7 joined to 8 by an oblique bar; the head with enormous frontal swelling   *Baljexifrons*

9. Fore wing with vein 3 and 5 from close to angle of cell. Hind wing with veins 4, 5 not approximated towards origin   *Pyrausta*

GENUS *PYRAUSTA (EUTECTONA)*

Pyrausta, Sclirank, Fauna JSoica, ii, 2, p. 163 (1802).

Syllythria, Hilbn. Verz. p. 349 (1818).

Haematia, Hiibn. Verz. p. 349.

Nascia, Curt. Brit. Ent. vi, p. 599 (1840).

Herbula, Guen. Delt. $ Pyr. p. 175 (1854).

Ebulea, Guen. Delt. $ Pyr. p. 357.

Gyptitia, Snell. Tijd. v. Ent. xxvi, p. 138 (1883).

Eclipsoides, Meyr. Trans. Ent. Soc. 1884, p. 343.

Paliga, Moore, Lep. Ceyl. iii, p. 350 (1885).

Protocolleiis, Meyr. Trans. Ent. Soc. 1888, p. 223.

Opsibotys, Warr. A. M. N. H. (6) vi, p. 474 (1890).

Sciorista, Warr. A. M. N. H. (6) vi, p. 475.

Micractis, Warr. A. M. N. H. (6) ix, p. 294 (1892).

Glauconoe, Warr. A. M. N. H. (6) ix, p. 296.

Aplographe, Warr. A. M. N. H. (6) ix, p. 301.

Crypsiptya, Meyr. Trans. Ent. Soc. 1894, p. 463.

Type, *P. cingulata,* Linn., from Europe.

Range. Universally distributed.

The genus is characterised by

1. Palpi porrect, triangularly scaled, the 3rd joint hidden by hair; maxillary palpi filiform; frons rounded.

2. Antennae not more than three-fourths length of fore wing and minutely ciliated.

3. Tibiae with the outer spurs short, the outer medial spur of hind tibiae not more than two-thirds the length of the inner spur.

4. Fore wing with veins 3 and 5 from close to angle of cell; 10 free or rarely anastomosing with 8, 9. Hind wing with veins 4, 5 not approximated towards origin; 6, 7 from upper angle, 7 anastomosing with 8.

From observations on material collected in China, the genus Eutectona is erected for the pyralid hitherto known as *Pyrausta machaeralis* (Wlk.) an important pest ofteak (*Tectona grandis*) in tropical Asia.

Eutecton machaeralis Walker (Figure 127)

Female

Body 21 mm long, 3 mm broad, wing expanse 33 mm.

Head

2 mm long, 1.4 mm broad; antenna 14 mm long, filiform yellowish, scape 0.6 mm, pedicel 0.4 mm, flagellum 13 mm.

Thorax

3 mm long, 2 mm broad, collar yellow. **Fore wing :** 16 mm long, 11mm broad, yellowish with zig-zag marking on dorsal side. **Hind wing :** 13 mm long, 10mm broad, yellowish. **Hind leg :** 14 mm long, coxa 4 mm, trochanter 0.4 mm, femur 5mm, tiba 4 mm, tarsus 2.6 mm.

Abdomen

15 mm long, 3 mm broad, yellowish.

Host Plant:

Tectona grandis

Holotype

Female, India, Satara, Koyananagar, VII 2010, coll., V. Y. Kadam, pinned insect in insect box and labeled as above.

Paratype

Female 10 male 5, Sex ratio (M :F) 1:2, coll. June to October, Koyanangar, V. Y. Kadam.

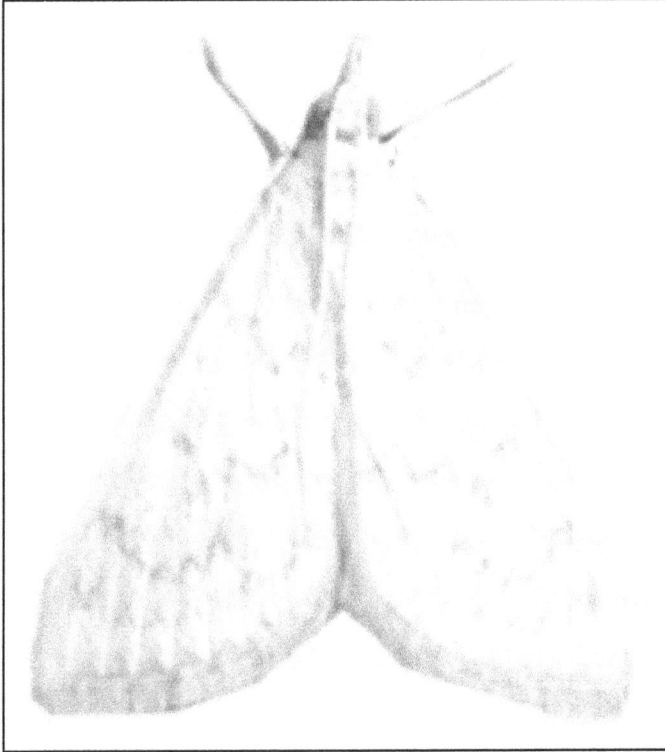

PLATE 36—Figure 127: *Eutecton machaeralis* **Walker.**

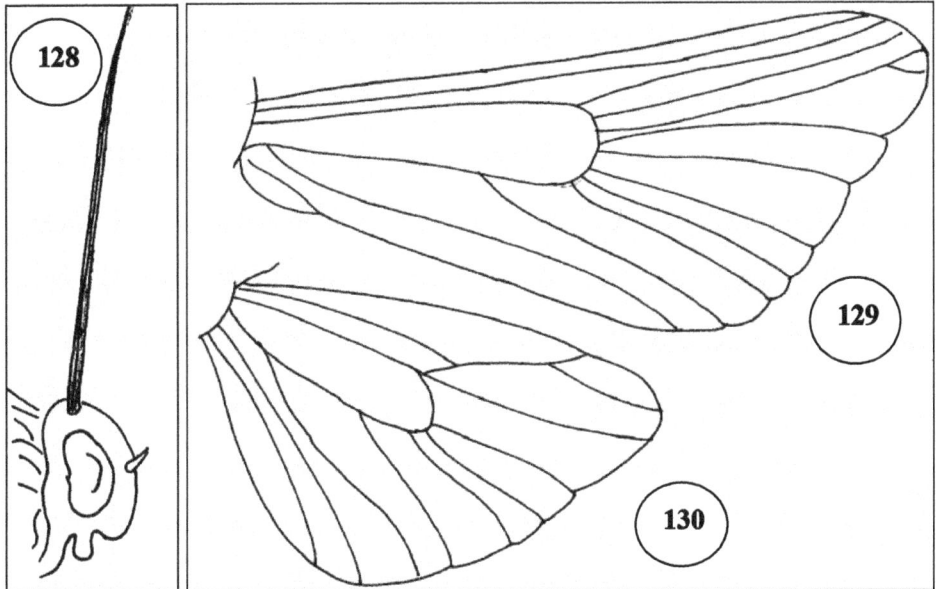

Figure 128: Head and Appendages.

Figure 129: Fore Wing Venation.

Figure 130: Hind Wing Venation.

Distribution

Through out India.

Control

Spray Rogor 0.03 per cent or Carbaryl 0.15 per cent, Malathion 0.03 per cent.

Remarks

According Hampson (1976) this species is *Eutectona machaeralis* in the text some additional characters are given for this since it is poorly described by Hampson.

5

Biology and Intrinsic Rate of Increase of Moths

Biology of *Eligma narcissus* Cram.

During present studies *E. narcissus* was found to be very serious pest of *A. excelsa* and therefore, detailed studies on biology was undertaken in laboratory at 25 ± 2°C, 65 ± 5 per cent R.H. and 12 hr photoperiod.

The initial culture of the pest was obtained by collecting larval stages from western Ghats of Maharashtra (Kolhapur and Satara district) during first fortnight of July, 2011 and reared in the laboratory in glass jars. Leaves of *A. excelsa* were provided as a food for the larvae and it was changed periodically till the pupation. Freshly emerged moths from the culture were sexed.

Mating

A newly emerged female and two male adults were released in glass jar covered with muslin cloth. The cotton swab soaked in five per cent sugar solution was kept suspended in the jar as food, which was changed daily. A total of ten such jars were kept under obserevations. Mating took place in the night.

Site of Oviposition

A female laid eggs on the surface of the leaves of *A. excelsa*.

Preoviposition, Oviposition and Postoviposition Periods

Preoviposition, oviposition and post oviposition periods in ten females were recorded separately. Result (Table 1) revealed that preovipositon period ranged from 2.0 to 4.0 days with average of 3.5 ± 0.71 days. In ten individuals, oviposition period ranges from 4.0 to 6.0 days with an average of 5.0 ± 0.67 days and post oviposition period ranged from 1.0 to 2.0 days with an average of 1.5 ± 0.53 days.

Fecundity

The results recorded (Table 1) indicated that the number of eggs laid by females ranged from 137 to 163 with average of 151.8 ± 8.58 days.

Table 1: Preoviposition, Oviposition, Postoviposition Periods and Fecundity of *E. narcissus*

Female Number	Preoviposition Period (Days)	Oviposition Period (Days)	Postovipostion Period (Days)	Fecundity
1	3	5	1	137
2	4	4	2	158
3	4	5	1	162
4	3	4	2	139
5	4	6	2	155
6	3	5	1	147
7	4	5	2	149
8	2	5	1	154
9	4	6	2	154
10	4	5	1	163
Mean	3.5 ± 0.71	5.0 ± 0.67	1.5 ± 0.53	151.8 ± 8.58
Range	2 to 4	4 to 6	1 to 2	137 to 163

Morphometrics of Eggs

Freshly laid eggs were spherical and pale yellow. In 20 individuals egg averaged 0.83 ± 0.024 mm in length and 0.66 ± 0.029 mm in breadth (Table 2).

Table 2: Morphometrics of Eggs of *E. narcissus*

Sl.No.	Length (mm)	Breadth (mm)
1.	0.87	0.67
2.	0.81	0.61
3.	0.83	0.70
4.	0.86	0.69
5.	0.83	0.62
6.	0.81	0.69
7.	0.83	0.67
8.	0.87	0.66
9.	0.84	0.65
10.	0.82	0.71

Contd...

Table 2–Contd...

Sl.No.	Length (mm)	Breadth (mm)
11.	0.85	0.62
12.	0.82	0.67
13.	0.86	0.64
14.	0.82	0.69
15.	0.84	0.63
16.	0.81	0.70
17.	0.81	0.66
18.	0.83	0.65
19.	0.79	0.67
20.	0.80	0.69
Mean	0.83 ± 0.024	0.66 ± 0.029
Range	0.79 to 0.87	0.61 to 0.71

Incubation Period and Hatching Percentage

The data on incubation period and hatching percentage were given in Table 3. Incubation period in 20 individuals ranged from 3 to 4 days (mean 3.42 ± 0.44 days). Maximum eggs hatched on 3rd day. Average egg hatching percentage was 75 ± 10.

Table 3: Incubation Period and Hatching Percentage of *E. narcissus*

Sl.No.	Date of Egg Laying	Eggs Observed	Egg Hatched	Incubation Period (days)	Per cent Hatch
1.	13.7.11	10	7	4.0	70
2.	13.7.11	10	6	4.0	60
3.	13.7.11	10	5	3.0	50
4.	13.7.11	10	7	3.0	70
5.	13.7.11	10	7	4.0	70
6.	13.7.11	10	7	3.0	70
7.	13.7.11	10	8	3.0	80
8.	14.7.11	10	8	3.5	80
9.	14.7.11	10	9	3.0	90
10.	14.7.11	10	8	4.0	80
11.	14.7.11	10	6	3.0	60
12.	14.7.11	10	8	3.0	80
13.	14.7.11	10	8	3.0	80
14.	14.7.11	10	9	4.0	90
15.	15.7.11	10	7	3.0	70

Contd...

Table 3–Contd...

Sl.No.	Date of Egg Laying	Eggs Observed	Egg Hatched	Incubation Period (days)	Per cent Hatch
16.	15.7.11	10	8	3.5	80
17	15.7.11	10	8	4.0	80
18	15.7.11	10	8	3.0	80
19	15.7.11	10	8	3.5	80
20	15.7.11	10	8	4.0	80
Mean		200	7.5 ± 1	3.42 ± 0.44	75 ± 10

Larval Developments

Observations on larval development are given in Table 4. The developmental period of larval instars ranged from 21 to 24 days with average of 22.15 ± 0.93 days. During this period a larva moulted 4 times. The instarwise measurements of head width, body length and body width are presented in Table 6.

Table 4: Development Period Required by Larval Instars of *E. narcissus*

Sl.No.	Instars					Total Period (Days)
	I	II	III	IV	V	
1.	3	5	4	4	5	21
2.	3	5	5	5	6	24
3.	3	5	4	5	5	22
4.	3	5	4	4	5	21
5.	3	5	4	5	5	22
6.	3	5	4	4	5	21
7.	3	5	4	5	5	22
8.	4	5	4	5	6	24
9.	3	4	5	5	5	22
10.	3	5	4	5	5	22
11.	3	4	5	5	5	22
12.	3	5	4	5	6	23
13.	3	5	4	5	5	22
14.	3	5	4	5	5	21
15.	3	5	4	5	6	22
16.	3	5	4	4	5	21
17.	3	4	5	5	5	23
18.	3	5	4	5	5	22
19.	3	5	4	5	5	22
20.	3	4	5	5	6	23
Mean	3.05 ± 0.22	4.8 ± 0.41	4.25 ± 0.44	4.8 ± 0.41	5.25 ± 0.51	22.15 ± 0.93
Range	3 to 4	4 to 5	4 to 5	4 to 5	5 to 6	21 to 24

First Instar

Newly hatched larva was tiny active and greenish yellow. Head capsule ranged from 0.29 mm to 0.39 mm (mean 0.34 mm ± 0.085 mm) in width (Table 6). Five pairs of prolegs were on 3rd, 4th, 5th, 6th and 10th abdominal segments. Larvae measured 2.9 mm to 3.2 mm in body length (mean 3.02 mm ± 0.078 mm) and 0.83 mm to 0.97 mm in width (mean 0.94 mm ± 0.038 mm) (Table 6). The instar lasted for 3 to 4 days (mean 3.05 ± 0.22 days) (Table 4).

Second Instar

This instar larva was very active and yellowish in colour. In 20 individuals head capsule ranged from 0.79 mm to 0.86 mm (mean 0.83 mm ± 0.019 mm) in width. Larvae ranged from 5.9 mm to 6.3 mm (mean 6.05 mm ± 0.095 mm) in length and 1.8 mm to 2.4 mm (mean 2.03 mm ± 0.19 mm) in width (Table 6). Instar lasted for 4 to 5 days (mean 4.8 ± 0.41 days) (Table 4).

Table 5: Developmental Period Required for different Stages of *E. narcissus*

Sl.No.	Developmental Period in Days			Adult Longevity in Days	
	Larva	Prepupa	Pupa	Male	Female
1.	21	1	5	—	9
2.	24	2	4	—	11
3.	22	1	6	—	10
4.	21	1	6	—	9
5.	22	1	7	10	—
6.	21	1	7	—	12
7.	22	1	6	—	9
8.	24	2	4	8	—
9.	22	1	5	—	10
10.	22	1	7	10	—
11.	22	1	5	6	—
12.	23	1	7	—	8
13.	22	1	5	—	12
14.	21	1	4	10	—
15.	23	2	7	—	10
16.	21	1	5	8	—
17.	23	1	6	—	8
18.	22	1	5	—	9
19.	22	1	7	9	—
20.	23	1	6	—	11
Mean	22.15 ± 0.93	1.15 ± 0.37	5.75 ± 0.067	8.7 ± 0.85	9.8 ± 1.03
Range	21 to 24	1 to 2	4 to 7	6 to 10	8 to 12

Table 6: Morphometrics of different Stages of *E. narcissus*

Sl.No.	Head Width (mm)					Body Width (mm)				
	I	II	III	IV	V	I	II	III	IV	V
1.	0.38	0.83	0.96	1.3	2.0	0.88	1.9	3.0	4.2	4.8
2.	0.35	0.81	0.91	1.5	2.1	0.89	1.9	3.1	4.0	4.7
3.	0.39	0.81	1.0	1.3	2.0	0.92	2.2	3.3	3.9	4.8
4.	0.32	0.83	0.98	1.4	2.0	0.95	2.1	3.3	3.9	4.8
5.	0.37	0.79	1.0	1.3	2.0	0.95	1.9	3.3	4.0	4.7
6.	0.29	0.83	0.95	1.3	2.1	0.83	2.0	3.1	4.0	4.8
7.	0.33	0.84	0.99	1.4	2.3	0.95	2.1	3.3	4.1	4.9
8.	0.37	0.82	0.92	1.3	2.1	0.97	2.4	3.2	4.2	4.7
9.	0.36	0.83	0.95	1.3	2.1	0.95	2.0	2.9	4.1	4.5
10.	0.31	0.83	1.0	1.4	1.9	0.97	2.2	3.3	4.0	4.5
11.	0.33	0.83	0.93	1.4	2.1	0.93	1.9	3.3	4.0	4.9
12.	0.31	0.82	0.98	1.3	2.3	0.95	2.0	3.4	4.2	4.5
13.	0.36	0.86	1.0	1.3	2.0	0.95	2.1	3.2	4.0	4.7
14.	0.29	0.86	0.98	1.5	2.0	0.98	1.8	3.3	4.1	4.9
15.	0.33	0.84	0.95	1.5	2.0	0.98	1.8	3.3	4.0	4.7
16.	0.35	0.85	1.0	1.4	2.1	0.93	2.3	2.9	4.0	4.7
17.	0.33	0.86	0.95	1.5	2.0	0.95	1.9	3.3	4.1	4.5
18.	0.37	0.81	1.0	1.3	2.0	0.97	2.0	2.9	4.0	5.1
19.	0.35	0.82	1.0	1.3	2.1	0.95	2.2	3.1	3.9	4.5
20.	0.35	0.80	0.95	1.5	2.1	0.95	2.0	3.1	3.9	4.5
Mean	0.34± 0.085	0.83± 0.019	0.97± 0.03	1.37± 0.85	2.06± 0.099	0.94± 0.038	2.03± 0.19	3.18 ± 0.16	4.03± 0.097	4.71± 0.17
Range	0.29 to.39	0.79 to.86	0.91 to 1.0	1.3 to 1.5	1.9 to 2.3	0.83 to 0.97	1.8 to 2.4	2.9 to3.4	3.9 to 4.2	4.5 to 5.1

Table 6 (Continued): Morphometrics of different Stages of *E. narcissus*

Sl.No.	Instars					Prepupa		Pupa	
	Body Length (mm)					Width	Length	Width	Length
	I	II	III	IV	V	(mm)	(mm)	(mm)	(mm)
1.	3.0	6.0	15.0	32	49	6.5	37	5.0	16.0
2.	3.1	6.1	15.0	31	53	6.0	37	5.1	16.0
3.	2.9	6.0	14.8	30	49	6.5	38	5.0	17.0
4.	3.0	6.0	15.0	30	49	6.1	37	5.1	17
5.	3.0	6.1	15.0	32	53	6.5	39	5.2	16
6.	3.0	6.1	15.0	30	50	6.1	37	5.0	16
7.	3.0	6.0	15.3	31	51	6.7	37	5.2	17
8.	3.0	5.9	15.0	30	52	6.3	39	5.0	16
9.	2.9	6.2	15.0	30	50	6.0	37	5.0	16
10.	3.0	6.0	15.2	32	51	6.3	37	5.1	16
11.	3.0	6.0	15.5	30	49	6.0	37	5.0	16
12.	3.1	6.3	15.0	32	53	5.9	38	5.0	17
13.	3.1	6.0	15.1	30	52	6.2	37	5.0	16
14.	3.1	6.0	15.3	32	50	6.1	39	5.0	15
15.	3.0	6.0	15.0	31	49	6.0	37	5.1	16
16.	2.9	6.1	15.1	31	50	6.3	39	5.0	16
17.	3.2	6.0	15.0	31	52	6.3	37	5.0	18
18.	3.0	6.0	15.1	32	53	6.3	39	5.0	17
19.	3.1	6.0	15.0	33	51	6.1	37	5.1	16
20.	3.1	6.2	15.0	30	50	6.0	37	5.1	17
Mean	3.02 ± 0.78	6.05 ± 0.095	15.07 ± 0.15	31 ± 0.97	50.8 ± 1.51	6.21 ± 0.22	37.6 ± 0.83	5.05 ± 0.69	16.35 ± 0.67
Range	2.9 to3.2	5.9 to 6.3	14.8 to15.5	30 to 33	49 to 53	5.9 to 6.7	37 to 39	5.0 to 5.2	15 to18

Table 6 (Continued): Morphometrics of different Stages of *E. narcissus*

Sl.No.	Adult					
	Male			Female		
	Body Length (mm)	Body Width (mm)	Wing Expansion (mm)	Body Length (mm)	Body Width (mm)	Wing Expansion (mm)
1.	20	4.0	62	23	4.0	66
2.	19	4.0	62	23	4.0	64
3.	21	4.1	61	23	4.1	66
4.	20	3.9	61	23	4.0	66
5.	21	4.0	58	21	4.0	66
6.	20	4.0	61	23	4.0	68
7.	20	4.0	61	21	4.0	68
8.	20	4.0	62	23	4.0	66
9.	19	4.0	62	23	4.0	66
10.	20	4.0	62	22	4.0	68
11.	20	4.1	62	23	4.1	66
12.	21	4.0	61	22	4.0	66
13.	20	4.0	59	22	4.1	66
14.	20	3.9	62	23	4.0	64
15.	19	4.0	62	22	4.0	66
16.	20	4.0	62	22	4.1	66
17.	20	4.0	62	22	4.0	66
18.	20	4.1	62	23	4.1	66
19.	19	3.9	61	23	4.0	68
20.	20	4.0	60	23	4.0	68
Mean	19.95 ± 0.6	4.0± 0.56	61.25 ± 1.12	22.5 ± 0.69	4.025 ± 0.044	66.3 ± 1.17
Range	19 to 21	3.9 to4.1	58 to 62	21 to 23	4.0 to 4.1	64 to 68

Third Instar

Larva was yellow in colour with black bands on the dorsal surface of the body and with white long hairs all over the body.

The head capsule in 20 individuals measured 0.91mm to 1.00 mm (mean 0.97 mm ± 0.03 mm) in width. Larvae measured from 14.8 mm to 15.5 mm (mean 15.07 mm ± 0.15 mm) in length and 2.9 mm to 3.4 mm (mean 3.18 mm ± 0.16 mm) in width (Table 6). The instar lasted for 4 to 5 days (mean 4.25 ± 0.44 days) (Table 4).

Fourth Instar

Larva was dark yellow with dark black bands on the dorsal surface of the body with white hairs all over the body.

The head capsules measured from 1.3 mm to 1.5 mm (mean 1.37 mm ± 0.085 mm) in width. The larvae were measured from 30 mm to 34 mm (mean 31 mm ± 0.97 mm) in length and 3.9 mm to 4.2 mm (mean 4.03 mm ± 0.097 mm) in width (Table 6). The instar lasted for 4 to 5 days (mean 4.8 ± 0.41 days) (Table 4).

Table 7: Statistics of Linear Regression Relationship of Larval Instars and Larval Head Width of *E. narcissus*

Sl.No.	Instar X	X^2	Head Width Y	Y^2	XY	Expected Values Y
1.	1	1	0.34	0.1156	0.34	0.318
2.	2	4	0.83	0.6869	1.66	0.716
3.	3	9	0.97	0.9409	2.91	1.114
4.	4	16	1.37	1.8769	5.48	1.512
5.	5	25	2.06	4.2436	10.3	1.91
Σ	3		1.114			

Table 8: Statistics of Linear Regression Relationship of Larval Instars and Larval Body Width of *E. narcissus*

Sl.No.	Instar X	X^2	Body Width Y	Y^2	XY	Expected Values Y
1.	1	1	0.94	0.8836	0.94	1.070
2.	2	4	2.03	4.1209	4.06	2.024
3.	3	9	3.18	10.1124	9.54	2.978
4.	4	16	4.03	16.2409	16.12	3.932
5.	5	25	4.71	22.1841	23.55	4.886
Σ	3		2.978			

Fifth Instar

The larva became elongated in this instar and colour becames dark yellow with black bands and body surface showed very long white hair. Head was black. Prolegs

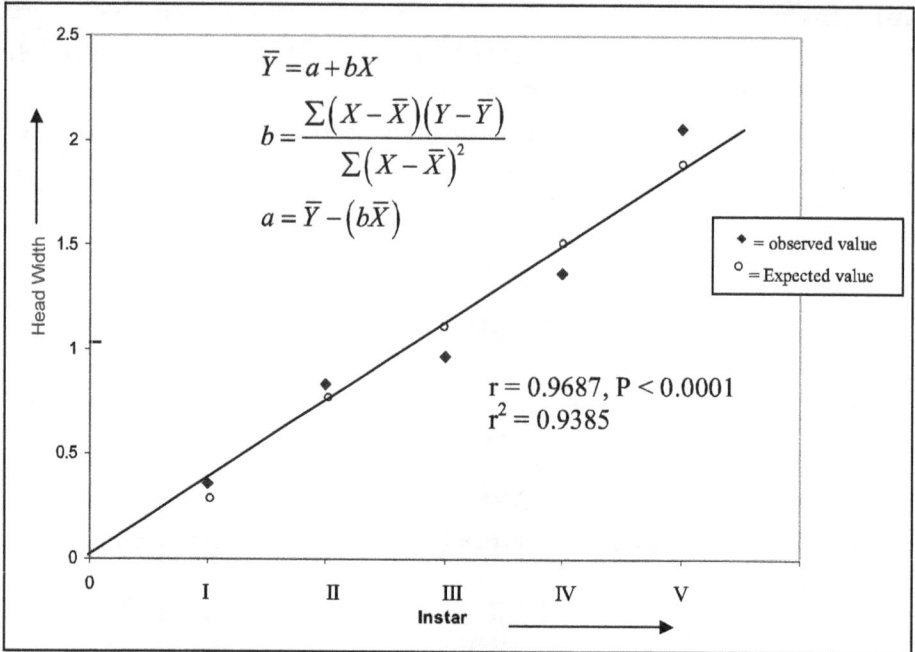

$$\overline{Y} = a + bX$$

$$b = \frac{\sum(X - \overline{X})(Y - \overline{Y})}{\sum(X - \overline{X})^2}$$

$$a = \overline{Y} - (b\overline{X})$$

◆ = observed value
○ = Expected value

$r = 0.9687, P < 0.0001$
$r^2 = 0.9385$

Figure 131: Relation between Larval Head Width and Larval Instars of *E. narcissus.*

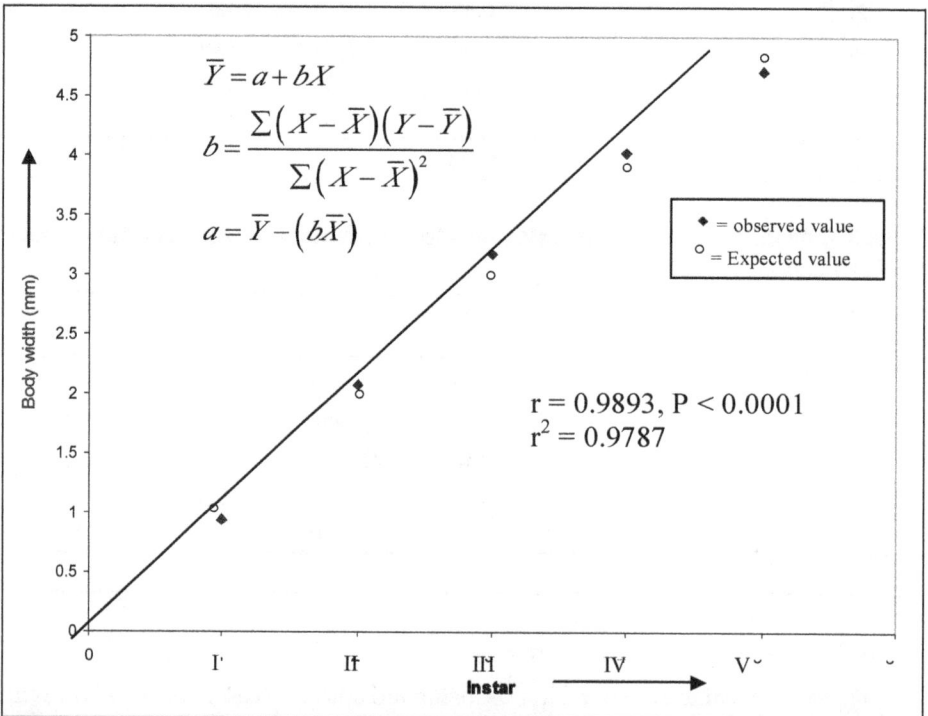

$$\overline{Y} = a + bX$$

$$b = \frac{\sum(X - \overline{X})(Y - \overline{Y})}{\sum(X - \overline{X})^2}$$

$$a = \overline{Y} - (b\overline{X})$$

◆ = observed value
○ = Expected value

$r = 0.9893, P < 0.0001$
$r^2 = 0.9787$

Figure 132: Relation between Larval Body Width and Larval Instars of *E. narcissus.*

were conspicuous. In 20 individuals head capsules measured from 1.9 mm to 2.3 mm (mean 2.06 mm ± 0.099 mm) in width (Table 6). In 20 individuals larvae ranged from 49 mm to 53 mm (mean 50.8 mm ± 1.51 mm) in body length and 4.5 mm to 5.1 mm (mean 4.71 mm ± 0.17 mm) in width (Table 6). Instar lasted for 5 to 6 days (mean 5.25 ± 0.51 days) (Table 4).

Instars of *E.narcissus* showed increased head capsule, body width and body length. The instars and their size was tested by applying regression analysis and found positively correlated and represented in Tables 7–9 and Figures 131–133.

Table 9: Statistics of Linear Regression Relationship of Larval Instars and Larval Body Length of *E. narcissus*

Sl.No.	Instar X	X^2	Body Length Y	Y^2	XY	Expected Values Y
1.	1	1	3.02	9.1204	3.02	−2.842
2.	2	4	6.05	36.6025	12.1	9.173
3.	3	9	15.07	227.1049	45.21	21.188
4.	4	16	31.00	961.00	124.00	33.203
5.	5	25	50.80	2580.00	254.00	45.218
Σ	3		21.188			

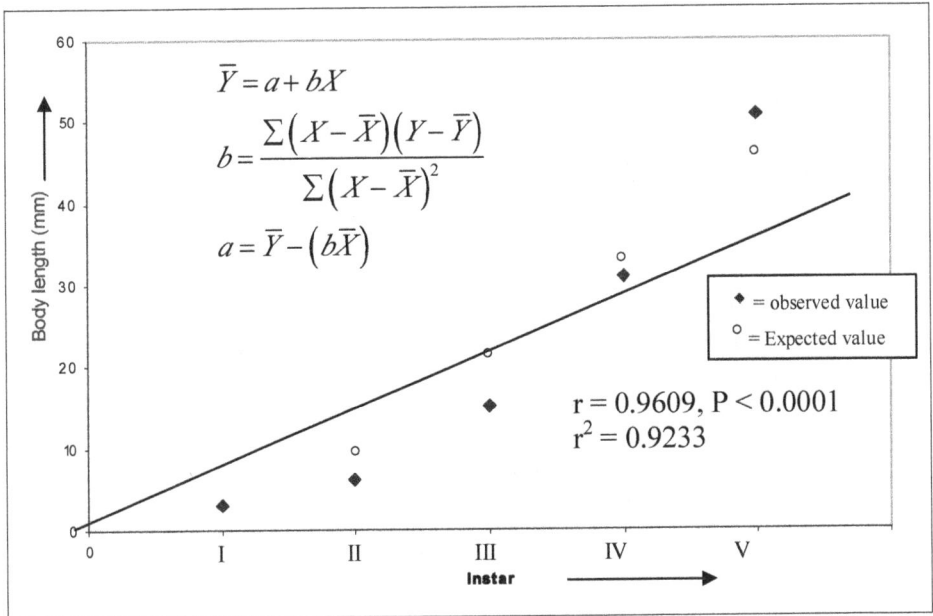

The plot shows the equations:

$$\overline{Y} = a + bX$$

$$b = \frac{\Sigma(X - \overline{X})(Y - \overline{Y})}{\Sigma(X - \overline{X})^2}$$

$$a = \overline{Y} - (b\overline{X})$$

◆ = observed value
○ = Expected value

r = 0.9609, P < 0.0001
$r^2 = 0.9233$

Figure 133: Relation between Larval Body Length and Larval Instars of *E. narcissus*.

Nature of Damage and Larval Behaviour

Newly hatched larvae were able to feed on both young tender or young leaves of *A. excelsa* and cause defoliation. The older larvae were able to fed on older and matured

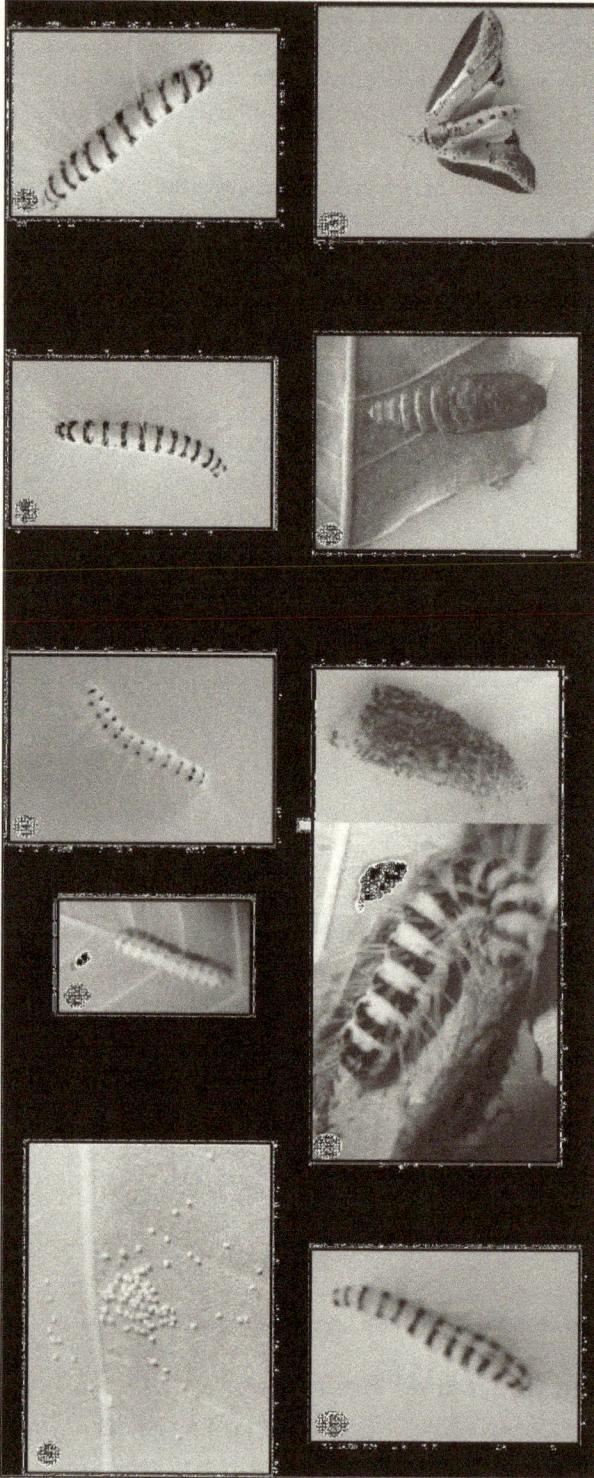

PLATE 37–Life Cycle of *Eligma narcissus*. Figure 143: Eggs; Figure 144: First instar larva; Figure 145: Second instar larva; Figure 146: Third instar larva; Figure 147: Fourth instar larva; Figure 148: Fifth instar larva; Figure 149: Prepupa; Figure 150: Pupa; Figure 151: Adult.

leaves of host plant. During heavy infestation as many as 20-40 larvae can be seen on rachis of compound leaf. When all the leaves are eaten up larvae may sometimes feed on green parts of the stem.

Prepupa

When larvae full grown became sluggish and stopped feeding and gradually shortened to the size of future pupa. In prepupa three distinct parts of body viz, head, thorax and abdomen were observed. Prepupal period lasted for 1 to 2 days (mean 1.15 ± 0.37days) (Table 5).

Pupa

In the laboratory larvae pupated on glass surface beneath the blotting paper in petri dish. Freshly formed pupa was reddish in colour and broader anteriorly and pointed posteriorly by ending in fine spine. In 20 individuals, pupal length ranged from 15 mm to 18 mm (mean 16.35 mm ± 0.67 mm) and width ranged from 5.0 mm to 5.2 mm (mean 5.05 mm ± 0.069 mm) (Table 6). Pupal stage lasted for 4 to 7 days (mean 5.75 ±0.67 days) (Table 5).

Adult

Abdomen of female was broader and bigger than that of male. In both sexes inner margin was light grey with 5 black spot at base. Outer side was dark black.

Female Adult

Female moth was medium sized and palpi were short and abdomen was bulging. Body length in 20 individuals ranged from 21 mm to 23 mm (mean 22.5 mm ± 0.69 mm) where as width ranged from 4.0 mm to 4.1 mm (mean 4.025 mm ± 0.044 mm), wing expansion ranged from 64 mm to 68 mm (mean 66.3 mm ± 1.17 mm) (Table 6). Longevity varied from 8 to 12 days (mean 9.8 ± 1.03 days) (Table 5).

Male Adult

Male moths were smaller than females and stoughtly built. Abdomen was slender, pointed towards anal end. Body length ranged from 19 mm to 21 mm (mean 19.95 mm ± 0.6 mm) while width ranged from 3.9 mm to 4.1 mm (mean 4 mm ± 0.56 mm), the wing expansion ranged from 58 mm to 62 mm (mean 61.25 mm ± 1.12 mm) (Table 6). Adult longevity varied from 6 to 10 days (mean 8.71 ± 0.85 days) (Table 5).

Life Cycle

Duration of egg, larval, prepual, pupal and adult stage varied from 3 to 4 days (mean 3.42 ± 0.44 days), 20 to 25 days (mean 22.15 ± 0.93 days), 1 to 2 days (1.15 ± 0.37 days), 4 to 7 days (mean 5.75 ± 0.67 days) and 8 to 12 days (9.8 ± 1.03 days) respectively. Larva completed its development in 22.15 ± 0.93 days by passing through five instars, the duration of which lasted for 3.05 ± 0.22, 4.8 ± 0.41, 4.25 ± 0.44, 4.8 ± 0.41, 5.25 ± 0.51 days respectively. The life cycle from egg to adult varied from 28 to 38 days (mean 32.47 ± 2.41 days). Period required for the completion of one generation varied from 36 to 50 days (mean 42.27 ± 3.44 days) (Table 10).

Sex Ratio

Sex ratio, male to female was 1:2.125 (Table 11).

Table 10: Life Cycle of *E. narcissus*

Stages	Duration in Days		
	Min.	Max.	Mean
Egg	3	4	3.42 ± 0.44
Larval Instars			
First instar	3	4	3.05 ± 0.22
Second instar	4	5	4.8 ± 0.41
Third instar	4	5	4.25 ± 0.44
Fourth instar	4	5	4.8 ± 0.41
Fifth instar	5	6	5.25 ± 0.51
Total larval period	20	25	22.15 ± 0.93
Prepupal period	1	2	1.15 ± 0.37
Pupal period	4	7	5.75 ± 0.67
Male	6	10	8.7 ± 0.85
Female	8	12	9.8 ± 1.03
Life cycle	28	38	32.47 ± 2.41
Period of generation	36	50	42.27 ± 3.44

Table 11: Sex Ratio of *E. narcissus*

Sl.No.	Adult Examined	Males	Females
1.	10	4	6
2.	10	2	8
3.	10	3	7
4.	10	4	6
5.	10	5	5
6.	10	4	6
7.	10	3	7
8.	10	2	8
9.	10	3	7
10.	10	2	8
Total	100	32	68

Sex ratio for male to female = 1:2.125

Biology of *Eutectona machaeralis* Walker

During present studies *E. machaeralis* was found to be very serious pest of *T. grandis* and therefore detailed studies on biology was undertaken in laboratory, at $25 \pm 2°C$, 65 ± 5 per cent R.H. and 12 hrs photoperiod.

The initial culture of the pest was obtained by collecting larval stages from western Ghats of Maharashtra (Kolhapur and Satara district) during first fortnight of September 2011 and reared in the laboratory in glass jars. Leaves of *T. grandis* were provided as a food for the larvae and it was changed periodically till the pupation. Freshly emerged moths from the culture were sexed.

Mating

Mating took place mostly in the dark hour and occasionally in the morning. A newly emerged female and two male adults were released in glass jar covered with muslin cloth. The cotton swab soaked in five per cent sugar solution was kept suspended in jar as food. A total of ten such jars were kept under obserevations.

Site of Oviposion

A female laid eggs singly on the undersurface of the leaves of *T. grandis*.

Preoviposition, Oviposition and Postoviposition Periods

Observation on ten females were recorded separately for preoviposition, oviposition and postoviposition periods. Result (Table 12) revealed that preovipositon period ranged from 2 to 3 days with an average of 2.6 ± 0.52 days.

Oviposition period ranged from 4 to 6 days with an average of 4.7 ± 0.67 days. Postoviposition period ranged from 1 to 2 days with an average of 1.4 ± 0.52 days.

Table 12: Preoviposition, Oviposition, Postoviposition Periods and Fecundity of *E. machaeralis*

Female Number	Preoviposition Period (Days)	Oviposition Period (Days)	Postovipostion Period (Days)	Fecundity
1	2	5	1	170
2	3	4	2	181
3	3	5	2	172
4	2	5	1	199
5	3	6	1	185
6	3	5	2	201
7	3	4	1	192
8	3	4	2	183
9	2	4	1	203
10	2	5	1	188
Mean	2.6 ± 0.52	4.7 ± 0.67	1.4 ± 0.52	187.4 ± 11.5
Range	2 to 3	4 to 6	1 to 2	170 to 203

Fecundity

Results indicated that the number of eggs laid by female ranged from 170 to 203 with an average of 187.4 ± 11.5 (Table 12) eggs.

Morphometrics of Eggs

Freshly laid eggs were spherical and pale yellow. In 20 individuals eggs averaged 0.69 mm ± 0.02 mm in length and 0.57 mm ± 0.022 mm in breadth (Table 13).

Table 13: Morphometrics of Eggs of *E. machaeralis*

Sl.No.	Length (mm)	Breadth (mm)
1.	0.69	0.60
2.	0.67	0.58
3.	0.65	0.55
4.	0.70	0.54
5.	0.71	0.59
6.	0.67	0.54
7.	0.66	0.58
8.	0.71	0.57
9.	0.69	0.55
10.	0.70	0.59
11.	0.72	0.57
12.	0.69	0.60
13.	0.70	0.61
14.	0.72	0.58
15.	0.69	0.60
16.	0.70	0.57
17.	0.68	0.58
18.	0.72	0.58
19.	0.71	0.55
20.	0.69	0.59
Mean	0.69 ± 0.02	0.57 ± 0.022
Range	0.65 to 0.72	0.54 to 0.61

Incubation Period and Hatching Percentage

The data on incubation period and hatching percentage of 20 individuals are given in Table 14. Incubation period ranged from 3 to 4 days (mean 3.25 ± 0.38 days). Average egg hatching percentage was 78.5 ± 9.88.

Larval Development

Observations on larval development are given in Table 15. The developmental period of larval instars ranged from 15 to 18 days with an average of 17.0 ± 1.12 days.

During this period, a larva moulted 4 times. The instarwise measurements on head width, body length and body width are presented in Table 17.

Table 14: Incubation Period and Hatching Percentage of *E. machaeralis*

Sl.No.	Date of Egg Laying	Eggs Observed	Egg Hatched	Incubation Period (days)	Per cent Hatch
1.	5.9.11	10	6	3.5	60
2.	5.9.11	10	6	3.0	60
3.	5.9.11	10	9	3.0	90
4.	5.9.11	10	9	3.0	90
5.	5.9.11	10	8	4.0	80
6.	5.9.11	10	8	3.0	80
7.	5.9.11	10	9	3.0	90
8.	6.9.11	10	8	3.0	80
9.	6.9.11	10	8	3.5	80
10.	6.9.11	10	8	3.5	80
11.	6.9.11	10	9	3.0	90
12.	6.9.11	10	9	3.0	90
13.	6.9.11	10	8	4.0	80
14.	6.9.11	10	9	3.0	90
15.	6.9.11	10	7	4.0	70
16.	7.9.11	10	7	3.0	70
17.	7.9.11	10	7	3.5	70
18.	7.9.11	10	7	3.0	70
19.	7.9.11	10	7	3.0	70
20.	7.9.11	10	8	3.0	80
Mean		200	7.85 ± 0.99	3.25 ± 0.38	78.5 ± 9.88

First Instar

Newly hatched larva was tiny active and greenish yellow. Head capsule ranged from 0.17 mm to 0.020 mm (mean 0.18 mm ± 0.011 mm) in width (Table 17). Five pairs of prolegs were on 3rd, 4th, 5th, 6th and 10th abdominal segments. Larvae measured 1.70 mm to 2.00 mm long (mean 1.8 mm ± 0.11mm) and 0.30 mm to 0.40 mm broad (mean 0.40 mm ± 0.22 mm) (Table 17). The instar lasted for 2 to 3 days (mean 2.65 ± 0.49 days) (Table 15).

Second Instar

Larva was very active and greenish in colour. Head capsule ranged 0.38 mm to 0.41 mm (mean 0.40 mm ± 0.008 mm) in width. Larvae ranged from 3.50 mm to 3.90 mm (mean 3.7 mm ± 0.14 mm) in length and 1.0 mm to 1.2 mm, (mean 1.1 mm ± 0.069 mm) in width (Table 17). This instar was lasted for 3 to 4 days (mean 3.6 ± 0.50 days) (Table 15).

Table 15: Development Period Required by Larval Instars of *E. machaeralis*

Sl.No.	Instars					Total Period (Days)
	I	*II*	*III*	*IV*	*V*	
1.	3.0	4.0	4.0	4.0	3.0	18.0
2.	3.0	3.0	3.0	3.0	4.0	16.0
3.	2.0	4.0	4.0	4.0	4.0	18.0
4.	3.0	3.0	3.0	4.0	3.0	15.0
5.	3.0	4.0	3.0	3.0	3.0	16.0
6.	3.0	3.0	3.0	4.0	5.0	18.0
7.	3.0	4.0	3.0	4.0	3.0	17.0
8.	2.0	4.0	3.0	3.0	3.0	15.0
9.	3.0	3.0	3.0	4.0	5.0	18.0
10.	2.0	4.0	4.0	4.0	3.0	17.0
11.	3.0	4.0	3.0	4.0	5.0	18.0
12.	3.0	4.0	4.0	3.0	4.0	18.0
13.	2.0	3.0	3.0	3.0	4.0	15.0
14.	3.0	4.0	3.0	4.0	4.0	18.0
15.	3.0	3.0	3.0	3.0	4.0	16.0
16.	3.0	4.0	4.0	3.0	3.0	17.0
17.	2.0	3.0	3.0	4.0	5.0	17.0
18.	3.0	4.0	3.0	4.0	4.0	18.0
19.	2.0	3.0	4.0	4.0	5.0	18.0
20.	2.0	4.0	3.0	4.0	4.0	17.0
Mean	2.65 ± 0.49	3.60 ± 0.50	3.30 ± 0.47	3.65 ± 0.49	3.90 ± 0.79	17.0 ± 1.12
Range	2 to 3	3.0 to 4.0	3.0 to 4.0	3.0 to 4.0	3.0 to 5.0	15.0 to 18.0

Third Instar

Third instar larva was green with light brown head bearing whitish setae on the body. Two pairs of black dots were visible on either side of the middorsal line of thoracic and abdominal segments.

The head capsules in 20 individuals measured from 0.60 mm to 0.63 mm (mean 0.62 mm ± 0.011 mm) in width. Larvae measured 10.9 mm to 11.4 mm (mean 11.2 mm ± 0.24 mm) in length and 2.0 mm to 2.2 mm (mean 2.09 mm ± 0.059 mm) in width (Table 17). The instar lasted for 3 to 4 days (mean 3.3 ± 0.47 days) (Table 15).

Fourth Instar

Larva was greenish brown with well developed head. The head capsule measured from 0.75 mm to 0.78 mm (mean 0.76 mm ± 0.009 mm) in width. Thoracic legs with claws at tip. The larvae were measured from 17.80 mm to 18.20 mm (mean 18 mm ±

0.13 mm) in length and 3.0 mm to 3.1 mm (mean 3.06 mm ± 0.051 mm) in width (Table 17). The instar lasted for 3 to 4 days (mean 3.65 ± 0.49 days) (Table 15).

Table 16: Developmental Period Required for different Stages of *E. machaeralis*

Sl.No.	Developmental Period in Days			Adult Longevity in Days	
	Larva	Prepupa	Pupa	Male	Female
1.	18.0	2.0	6.0	—	8
2.	16.0	2.0	6.0	5	—
3.	18.0	1.0	5.0	—	9
4.	15.0	1.0	6.0	6	—
5.	16.0	2.0	5.0	—	10
6.	18.0	1.0	5.0	—	8
7.	17.0	1.0	6.0	—	10
8.	17.0	2.0	6.0	6	—
9.	18.0	2.0	6.0	—	10
10.	17.0	1.0	5.0	6	—
11.	18.0	2.0	6.0	5	—
12.	18.0	1.0	6.0	5	—
13.	15.0	1.0	6.0	—	9
14.	18.0	2.0	5.0	5	—
15.	16.0	1.0	6.0	—	9
16.	17.0	1.0	6.0	4	—
17.	17.0	2.0	6.0	—	7
18.	18.0	2.0	6.0	—	8
19.	18.0	1.0	6.0	—	9
20.	17.0	1.0	5.0	—	8
Mean	17.0 ± 1.12	1.45 ± 0.51	5.70 ± 0.47	5.25 ± 0.75	8.75 ± 0.88
Range	15.0 to 18.0	1.0 to 2.0	5.0 to 6.0	4.0 to 6.0	7.0 to 10.0

Fifth Instar

The larva changed from greenish to brown. Two longitudinal bands observed on lateral side of thoraracic and abdominal terga. Head was brown, prolegs were conspicuous. Mouth parts were well developed. Head capsules were 1.1mm to 1.4 mm (mean 1.25 mm ± 0.089 mm) in width (Table 17). Larvae were 21.7 mm to 22.3 mm (mean 22 mm ± 0.21 mm) in length and 3.3 mm to 3.5 mm (mean 3.4 mm ± 0.079 mm) in width (Table 17). Instar lasted for 3 to 5 days (mean 3.9 ± 0.79 days) (Table 15).

Instars of *E.machaeralis* showed increased head capsule, body width and body length. The instars and their size was tested by applying regression analysis and found positively correlated and represented in Tables 18–20 and Figures 134–136.

Table 17: Morphometrics of different Stages of *E. machaeralis*

Sl.No.	Larval									
	Head Width (mm)					Body Width (mm)				
	I	II	III	IV	V	I	II	III	IV	V
1.	0.17	0.41	0.60	0.75	1.1	0.4	1.0	2.0	3.0	3.3
2.	0.18	0.40	0.61	0.76	1.3	0.4	1.1	2.1	3.1	3.4
3.	0.20	0.39	0.62	0.77	1.4	0.4	1.1	2.1	3.1	3.5
4.	0.17	0.40	0.63	0.77	1.3	0.4	1.0	2.0	3.0	3.3
5.	0.18	0.40	0.62	0.76	1.2	0.4	1.1	2.1	3.1	3.4
6.	0.20	0.39	0.62	0.76	1.2	0.4	1.2	2.1	3.1	3.5
7.	0.20	0.41	0.63	0.78	1.3	0.4	1.2	2.1	3.1	3.5
8.	0.18	0.38	0.61	0.76	1.3	0.4	1.2	2.1	3.1	3.4
9.	0.17	0.41	0.62	0.75	1.1	0.3	1.1	2.0	3.0	3.3
10.	0.19	0.40	0.63	0.78	1.3	0.4	1.2	2.2	3.1	3.5
11.	0.18	0.40	0.61	0.76	1.3	0.4	1.2	2.1	3.0	3.3
12.	0.20	0.41	0.63	0.76	1.2	0.4	1.1	2.2	3.1	3.5
13.	0.18	0.40	0.61	0.76	1.3	0.4	1.1	2.1	3.0	3.3
14.	0.17	0.41	0.62	0.76	1.1	0.4	1.0	2.0	3.0	3.3
15.	0.20	0.39	0.62	0.77	1.4	0.4	1.0	2.0	3.0	3.4
16.	0.18	0.40	0.62	0.76	1.2	0.4	1.1	2.1	3.0	3.4
17.	0.20	0.41	0.63	0.78	1.3	0.4	1.1	2.1	3.1	3.5
18.	0.18	0.40	0.62	0.76	1.2	0.4	1.1	2.1	3.0	3.4
19.	0.18	0.40	0.62	0.76	1.2	0.4	1.1	2.1	3.1	3.4
20.	0.18	0.40	0.63	0.77	1.3	0.4	1.1	2.1	3.1	3.4
Mean	0.18 ± 0.11	0.40 ± 0.008	0.62 ± 0.011	0.76 ± 0.009	1.25 ± 0.089	0.4 ± 0.022	1.1 ± 0.69	2.09 ± 0.059	3.06 ± 0.51	3.4 ± 0.79
Range	0.17 to 0.20	0.38 to 0.41	0.60to0.63	0.75 to 0.78	1.1 to 1.4	0.3 to 0.4	1.0 to 1.2	2.0 to 2.2	3.0 to 3.1	3.3 to 3.5

Table 17 (Continued): Morphometrics of different Stages of E. machaeralis

Sl.No.	Instars Body Length (mm)					Prepupa		Pupa	
	I (mm)	II	III	IV	V	Width (mm)	Length (mm)	Width (mm)	Length (mm)
1.	1.9	3.7	11.1	18.0	22.0	4.5	17.1	2.3	11.9
2.	1.8	3.6	11.0	18.1	21.8	4.4	17.0	2.2	11.8
3.	1.7	3.7	11.1	18.0	22.0	4.5	17.3	2.2	11.8
4.	1.9	3.7	11.2	18.2	22.0	4.5	17.3	2.2	11.9
5.	1.8	3.6	11.0	17.9	21.8	4.4	17.0	2.2	11.8
6.	1.7	3.5	11.1	17.8	22.2	4.3	16.9	2.1	11.7
7.	1.9	3.7	11.2	18.0	22.0	4.5	17.1	2.2	12.0
8.	1.7	3.9	11.4	18.2	22.3	4.6	17.3	2.1	12.2
9.	1.7	3.5	11.0	17.8	21.7	4.6	17.0	2.1	11.7
10.	1.7	3.9	11.4	17.8	21.7	4.6	11.3	2.1	11.7
11.	1.8	3.6	11.2	18.0	22.3	4.5	17.1	2.2	12.0
12.	1.9	3.8	11.3	18.1	22.1	4.6	17.2	2.3	12.1
13.	1.7	3.9	11.4	18.2	22.3	4.6	17.3	2.4	12.2
14.	1.7	3.9	11.4	17.8	22.2	4.6	16.9	2.1	11.7
15.	1.8	3.6	11.0	17.9	21.8	4.5	17.0	2.2	11.8
16.	1.8	3.6	11.0	17.9	21.8	4.4	17.0	2.2	11.8
17.	1.7	3.5	10.4	18.0	21.7	4.3	16.9	2.1	11.7
18.	1.9	3.7	11.2	18.0	22.0	4.5	17.1	2.2	12.0
19.	2.0	3.9	11.4	18.2	22.3	4.6	17.3	2.4	12.2
20.	1.9	3.7	11.2	18.0	22.0	4.5	17.1	2.2	12.0
Mean	1.8 ± 0.11	3.7 ± 0.14	11.2 ± 0.24	18.0 ± 0.13	22.0± 0.21	4.5 ± 0.11	17.0 ± 0.14	2.2 ± 0.99	11.9 ± 0.18
Range	1.7to2.0	3.5to3.9	10.9to11.4	17.8to18.2	21.7 to 22.3	4.3to4.6	16.9to17.3	2.1to2.4	11.7to12.2

Table 17 (Continued): Morphometrics of different Stages of *E. machaeralis*

Sl.No.	Adult					
	Male			Female		
	Body Length (mm)	Body Width (mm)	Wing Expansion (mm)	Body Length (mm)	Body Width (mm)	Wing Expansion (mm)
1.	12.0	1.7	19.0	9.1	2.3	20.0
2.	12.1	1.8	19.2	9.2	2.4	20.2
3.	11.9	1.6	18.7	8.8	2.2	18.2
4.	12.1	1.8	19.2	9.2	2.4	20.1
5.	12.1	1.8	19.2	9.2	2.4	20.1
6.	10.7	1.5	17.2	9.0	2.3	18.3
7.	11.9	1.8	18.7	8.8	2.2	18.2
8.	12.0	1.7	19.0	9.1	2.3	20.0
9.	10.8	1.8	17.3	9.1	2.4	18.3
10.	11.7	1.5	17.2	9.0	2.3	18.1
11.	11.9	1.6	18.7	8.8	2.2	19.1
12.	12.1	1.8	19.2	9.2	2.4	20.1
13.	12.0	1.7	19.0	9.1	2.3	20.0
14.	12.0	1.7	19.0	9.1	2.3	20.0
15.	11.9	1.6	18.7	8.8	2.2	18.2
16.	10.8	1.8	17.3	9.1	2.4	18.3
17.	12.0	1.7	19.0	9.1	2.3	20.0
18.	12.0	1.7	19.0	9.1	2.3	20.0
19.	12.1	1.8	19.2	9.2	2.4	20.2
20.	11.9	1.6	18.7	8.8	2.2	18.6
Mean	11.8 ± 0.52	1.7 ± 0.98	18.6 ± 0.73	9.04 ± 0.15	2.3 ± 0.79	19.3 ± 0.92
Range	10.7 to 12.1	1.5 to 1.8	17.2 to 19.2	8.8 to 9.2	2.2 to 2.4	18.1 to 20.2

Table 18: Statistics of Linear Regression Relationship of Larval Instars and Larval Head Width of *E. machaeralis*

Sl.No.	Instar X	X^2	Head Width Y	Y^2	XY	Expected Values Y
1.	1	1	0.18	0.0324	0.18	0.0192
2.	2	4	0.40	0.16	0.80	0.3306
3.	3	9	0.62	0.3844	1.86	0.642
4.	4	16	0.76	0.5776	3.04	0.9534
5.	5	25	1.25	1.5625	6.25	1.2648
Σ		3		0.642		

Table 19: Statistics of Linear Regression Relationship of Larval Instars and Larval Body Width of *E. machaeralis*

Sl.No.	Instar X	X^2	Body Width Y	Y^2	XY	Expected Values Y
1.	1	1	0.4	0.16	0.4	0.418
2.	2	4	1.1	1.21	2.2	1.214
3.	3	9	2.09	4.3681	6.27	2.388
4.	4	16	3.06	9.3636	12.24	3.184
5.	5	25	3.4	11.56	17.00	3.98
Σ		3		2.01		

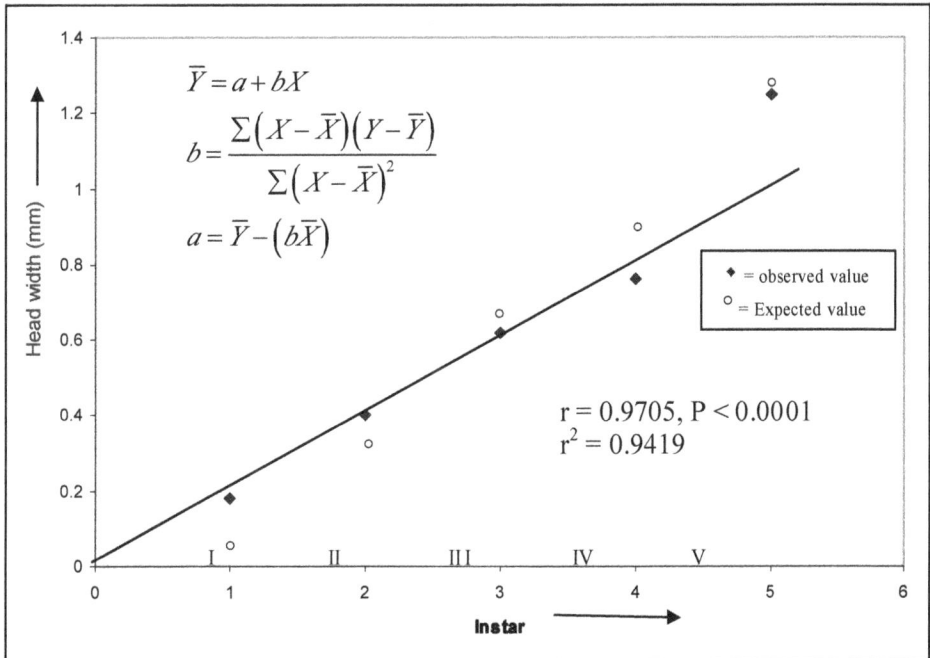

$$\overline{Y} = a + bX$$

$$b = \frac{\Sigma(X - \overline{X})(Y - \overline{Y})}{\Sigma(X - \overline{X})^2}$$

$$a = \overline{Y} - (b\overline{X})$$

$\blacklozenge$ = observed value
$\circ$ = Expected value

$r = 0.9705, P < 0.0001$
$r^2 = 0.9419$

Figure 134: Relation between Larval Head Width and Larval Instars of *E. machaeralis*.

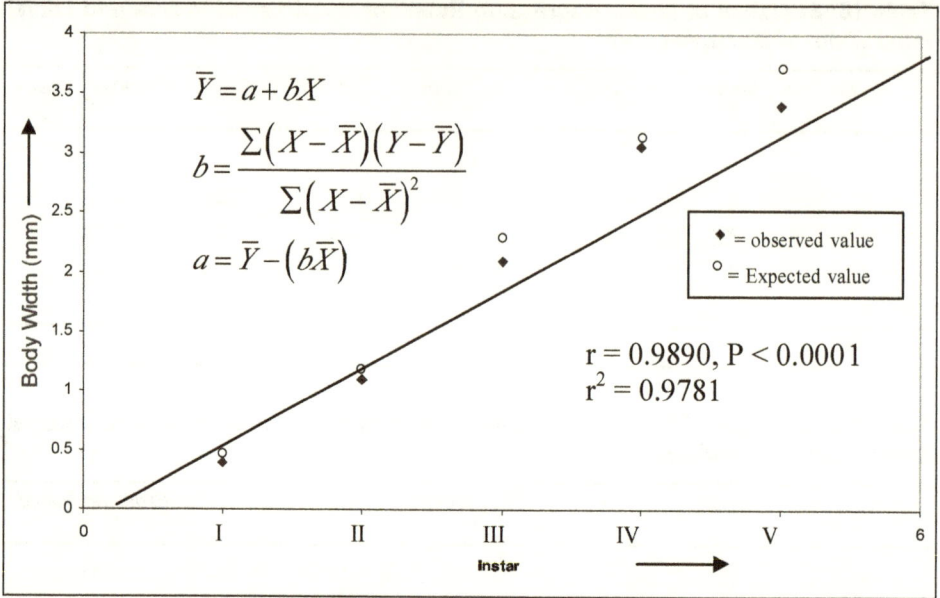

Figure 135: Relation between Larval Body Width and Larval Instars of *E. machaeralis.*

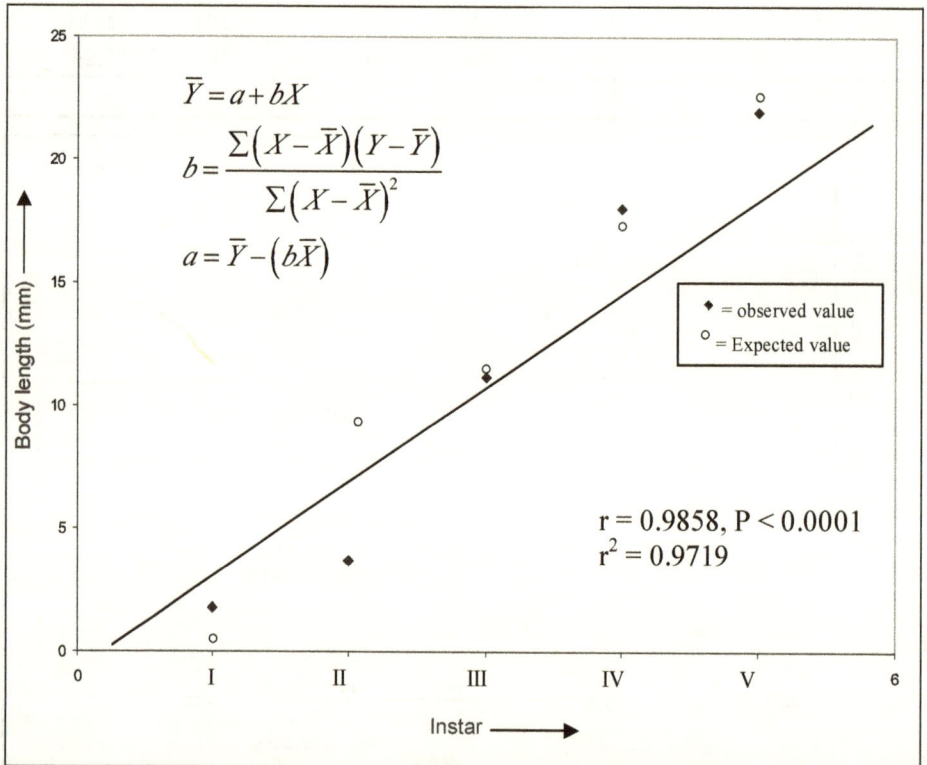

Figure 136: Relation between Larval Body Length and Larval Instars of *E. machaeralis.*

Table 20 : Statistics of linear regression relationship of larval instars and larval body length of *E. machaeralis*

Sl.No.	Instar X	X^2	Body Length Y	Y^2	XY	Expected Values Y
1.	1	1	1.8	3.25	1.8	0.40
2.	2	4	3.7	13.69	7.4	5.87
3.	3	9	11.2	125.44	33.6	11.34
4.	4	16	18.00	324.00	72.00	16.81
5.	5	25	22.00	484	110.00	22.28
Σ	3		11.34			

Nature of Damage and Larval Behaviour

Newly hatched larva was very active and fed on green matter under leaf. Second instar larva consumed tissues between the network of veins. Later instars migrated from one leaf to another and fed on green matter thereby skeletonising the leaf which turned brown. Tender leaves more badly damaged than older leaves. Moulting took place under web.

Prepupa

The full grown larva was sluggish and stopped its feeding and changed from greenish brown to purple. The larva gradually shortened to the size of future pupa. A thick and opaque shelter web was spun by the larva. Prepupal period lasted for 1.0 to 2 days (mean 1.45 ± 0.51 days) (Table 16).

Pupa

The larvae pupated on either lower or upper surface of leaf. The larva also pupated on glass surface beneath the blotting paper. Freshly formed pupa was reddish brown, gradually turned dark brown. In 10 individuals pupal length ranged from 11.7 mm to 12. 2 mm (mean 11.9 mm ± 0.18 mm) and width ranged from 2.1 mm to 2.4 mm (mean 2.2 mm ± 0.099 mm) (Table 17). Pupal stage lasted for 5 to 6 days (mean 5.7 ± 0.47days) (Table 16).

Adult

Upon eclosion, wings expanded in about ten minutes and moth was ready to fly. Abdomen of female broader and bigger than that of male. In both sexes fore wings were bright yellow with pink transverse zigzag marking while hind wings were pale yellow with reddish marginal lines and antennae were setaceous.

Female Adult

Female moth was medium sized and strongly built. Tarsal spurs on segment were pointed. Compound eyes were reddish brown. The abdomen of female was bulging. In 20 individuals body length ranged from 8.8 mm to 9.2 mm (mean 9.04 mm ± 0.15 mm) whereas width ranged from 2.2 mm to 2.4 mm (mean 2.3 mm ± 0.079 mm).

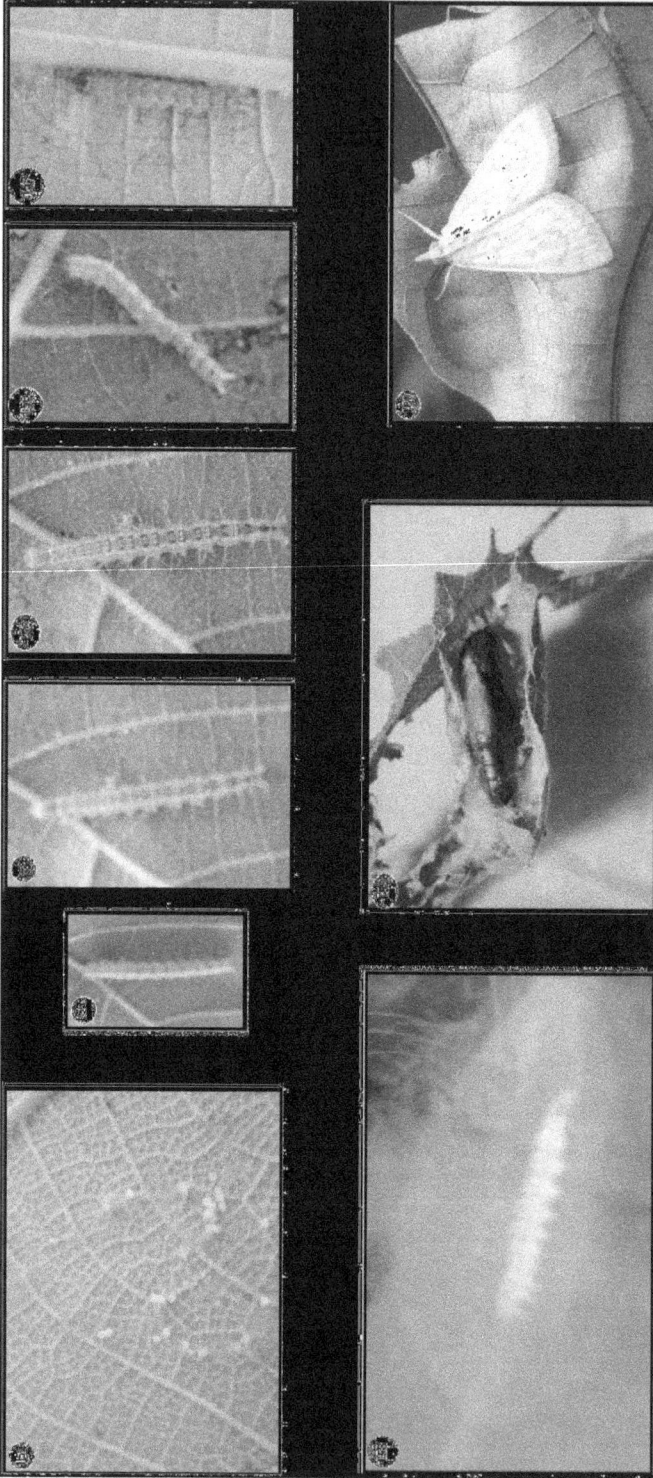

PLATE 38—Life Cycle of *Eutectona machaeralis*. Figure 152: Eggs; Figure 153: First instar larva; Figure 154: Second instar larva; Figure 155: Third instar larva; Figure 156: Fourth instar larva; Figure 157: Fifth instar larva; Figure 158: Prepupa; Figure 159: Pupa; Figure 160: Adult.

Wing expansion ranged from 18.1 mm to 20.2 mm (mean 19.3 mm ± 0.93 mm) (Table 17). Adult longevity varied from 7 to 10 days (mean 8.75 ± 0.88 days) (Table 16).

Male Adult

Moth was medium sized, stoutly built but smaller than females. In male abdomen was slender and pointed towards anal end. In 20 individuals body length ranged from 10.7 mm to 12. 1 mm (mean 11.8 mm ± 0.52mm) while the width ranged from 1.5 mm to 1.8 mm (mean 1.7 mm ± 0.098 mm). The wing expansion ranged from 17.2 mm to 19.2 mm (mean 18.6 mm ± 0.73 mm) (Table 17). Adult longevity varied from 4 to 6 days (mean 5.25 ± 0.75 days) (Table 16).

Life Cycle

The duration of egg, larval, prepupal, pupal and adult stages varied from 3 to 4 days (mean 3.25 ± 0.38 days), 15 to 18 days (mean 17.0 ± 1.12 days), 1 to 2 days (mean 1.45 ± 0.51 days), 5 to 6 days (mean 5.7 ± 0.47 days) and 7 to 10 days (mean 8.75 mm ± 0.88 mm) respectively (Table 21). Larva completed its development in 17.0 ± 1.12 days by passing through five instars, the duration of which lasted for 2.65 ± 0.49, 3.60 ± 0.50, 3.30 ±0.47, 3.65 ± 49 and 3.90 ± 0.79 days respectively. The life cycle from egg to adult varied from 24.00 to 30.00 (mean 27.40days). Period required for the completion of one generation varied from 31.00 to 40.00 (mean 36.15 ± 3.36 days) (Table 21).

Table 21: Life Cycle of *E. machaeralis*

Stages	Duration in Days		
	Min.	Max.	Mean
Egg	3.0	4.0	3.25 ± 0.38
Larval Instars			
First instar	2.0	3.0	2.65 ± 0.49
Second instar	3.0	4.0	3.60 ± 0.50
Third instar	3.0	4.0	3.30 ± 0.47
Fourth instar	3.0	4.0	3.65 ± 0.59
Fifth instar	3.0	5.0	3.90 ± 0.79
Total larval period	15.0	18.0	17.0 ± 1.12
Pre pupal period	1.0	2.0	1.45 ± 0.51
Pupal period	5.0	6.0	5.70 ± 0.47
Male	4.0	6.0	5.25 ± 0.75
Female	7.0	10.0	8.75 ± 0.88
Life cycle	24.0	30.0	27.40 ± 2.48
Period of generation.	31.0	40.0	36.15 ± 3.36

Sex Ratio

Sex ratio, male to female averaged 1:1.5 (Table 22).

Table 22: Sex Ratio of *E. machaeralis*

Sl.No.	Adult Examined	Males	Females
1	10	4	6
2	10	5	5
3	10	4	6
4	10	3	7
5	10	4	6
6	10	5	5
7	10	4	6
8	10	3	7
9	10	4	6
10	10	4	6
Total	100	40	60

Sex ratio, male to female = 1:1.5

Biology of *Hypsa producta* Moore

During present studies *H. producta* was found to be very serious pest of *F. glomerata* and therefore, detailed studies on biology was undertaken in laboratory at 25± 2°C, 65 ± 5 per cent R.H. and 12 hrs photoperiod.

The initial culture of the pest was obtained by collecting larval stages from western Ghats of Maharashtra (Kolhapur and Satara district) during first fortnight of November 2010 and reared in the laboratory in glass jars. Leaves of *F. glomerata* were provided as a food for the larvae and it was changed periodically till the pupation. Freshly emerged moths from the culture were sexed.

Mating

A newly emerged female and two male adults were released in glass jar covered with muslin cloth. The cotton swab soaked in five percent sugar solution was kept suspended in jar as food. A total of ten such jars were kept under observations.

Site of Oviposition

A female laid eggs on the surface of the leaves of *F.glomerata*.

Preoviposition, Oviposiion and Postovipostion Period

Observations on ten females were recorded separately for preoviposition, oviposition and postoviposition period. Table 23 revealed that preoviposition period ranged from 3.0 to 4.0 days with an average of 3.4 ± 0.52 days. Oviposition period ranged from 3 to 5 days with an average of 3.8 ± 0.79 days. Postoviposition period ranged from 2 to 3 days with an average of 2.4 ± 0.52 days.

Fecundity

Results recorded in Table 23 indicated that the number of eggs laid by female ranged from 147 to 169 with an average of 157.9 ± 6.89.

Table 23: Preoviposition, Oviposition, Postoviposition Periods and Fecundity of
H. producta

Female Number	Preoviposition Period (Days)	Oviposition Period (Days)	Postovipostion Period (Days)	Fecundity
1	3	3	3	157
2	3	4	2	152
3	3	4	2	160
4	3	5	2	164
5	4	4	2	151
6	3	3	3	147
7	4	5	3	169
8	3	3	2	165
9	4	4	3	155
10	4	3	2	159
Mean	3.4 ± 0.52	3.8 ± 0.79	2.4 ± 0.52	157.9 ± 6.89
Range	3 to 4	3 to 5	2 to 3	147 to 169

Table 24: Morphometrics of Eggs of *H. producta*

Sl.No.	Length (mm)	Breadth (mm)
1.	0.79	0.65
2.	0.83	0.69
3.	0.82	0.69
4.	0.82	0.69
5.	0.85	0.70
6.	0.81	0.68
7.	0.80	0.68
8.	0.83	0.69
9.	0.81	0.68
10.	0.83	0.69
11.	0.83	0.69
12.	0.81	0.66
13.	0.81	0.66
14.	0.83	0.69
15.	0.85	0.70
16.	0.85	0.70
17.	0.79	0.64
18.	0.80	0.65
19.	0.83	0.69
20.	0.83	0.68
Mean	0.82 ± 0.018	0.68 ± 0.18
Range	0.79 to 0.85	0.64 to 0.70

Morphometry of Eggs

Freshly laid eggs were spherical and measured 0.82 mm ± 0.018 mm in length and 0.68 ± 0.18 mm in breadth (Table 24).

Incubation Period and Hatching Percentage

The data on incubation period and hatching percentage is given in Table 25. Incubation period in 20 individuals ranged from 2.5 to 3.0 days (mean 3.0 ± 0.23 days). Average egg hatching percentage was 75 ± 6.07.

Table 25: Incubation Period and Hatching Percentage of *H. producta*

Sl.No.	Date of Egg Laying	Eggs Observed	Egg Hatched	Incubation Period (days)	Per cent Hatch
1.	3.11.2010	10	7	2.5	70
2.	3.11.2010	10	8	3.0	80
3.	3.11.2010	10	7	3.0	70
4.	3.11.2010	10	9	3.0	90
5.	3.11.2010	10	8	3.0	80
6.	3.11.2010	10	8	2.5	80
7.	3.11.2010	10	7	3.5	70
8.	3.11.2010	10	7	3.0	70
9.	3.11.2010	10	7	3.0	70
10.	3.11.2010	10	7	3.0	70
11.	3.11.2010	10	7	3.0	70
12.	4.11.2010	10	8	3.0	80
13.	4.11.2010	10	8	3.0	80
14.	4.11.2010	10	8	3.5	80
15.	4.11.2010	10	7	3.0	70
16.	5.11.2010	10	7	3.0	70
17.	5.11.2010	10	8	3.0	80
18.	5.11.2010	10	7	3.0	70
19.	5.11.2010	10	8	3.0	80
20.	5.11.2010	10	7	3.0	70
Mean		200	7.50± 0.61	3.0 ± 0.23	75 ± 6.07

Larval Development

Observations on larval development are given in Table 26. The development period of larval instars ranged from 18 to 21 days with average of 19.1 ± 1.36 days. During this period larva moulted four times. The instar wise measurements of head width, body length and body width are presented in Table 28.

Table 26: Developmental Period Required by Larval Instars of *H. producta*

Sl.No.	Instars					Total Period (Days)
	I	*II*	*III*	*IV*	*V*	
1.	2	4	5	4	5	20
2.	3	4	4	5	5	21
3.	2	4	4	4	5	19
4.	2	4	4	4	4	18
5.	3	4	5	4	5	21
6.	2	3	4	4	5	18
7.	3	4	5	4	5	21
8.	2	3	4	4	5	18
9.	3	4	5	4	5	21
10.	2	4	4	4	4	18
11.	2	3	4	4	5	18
12.	2	3	4	4	5	18
13.	2	3	4	4	5	18
14.	2	3	4	5	5	19
15.	2	3	4	5	4	18
16.	2	4	3	4	5	18
17.	2	4	3	4	4	17
18.	3	4	5	4	5	21
19.	2	3	4	5	5	19
20.	3	4	5	4	5	21
Mean	2.3 ± 0.44	3.6 ± 0.50	4.2 ± 0.62	4.2 ± 0.41	4.8 + 0.41	19.1 ±1.36
Range	2 to 3	3 to 4	3 to 5	4 to 5	4 to 5	18 to 21

First Instar

Newly hatched larva was active and black in colour. Head capsules ranged from 0.43 mm to 0.47 mm (mean 0.45 mm ± 0.016 mm) in width (Table 28). Five pairs of proleges were on 3rd, 4th, 5th, 6th and 10th abdominal segments. Larvae measured 2.7 mm to 3.1 mm in body length (mean 2.9 mm± 0.13 mm) and 0.92 mm to 0.95 mm (mean 0.93 ± 0.012 mm) in width (Table 28). The instar lasted for 2 to 3 days (mean 2.3 ± 0.44 days) (Table 32).

Second Instar

This instar larva was very active blackish in colour. In 20 individuals head capsule ranged from 0.93 mm to 0.97 mm (mean 0.95 mm ± 0.010 mm) in width. Larvae ranged from 5 mm to 5.3 mm (mean 5.2 mm ± 0.097 mm) in length and 2.3 mm to 2.5 mm. (mean 2.4 mm ± 0.14 mm) in body width (Table 28). The instar lasted for 3 to 4 days (mean 3.6 ± 0.50 days) (Table 26).

Table 27: Developmental Period Required for different Stages of *H. producta*

Sl.No.	Developmental Period in Days			Adult Longevity in Days	
	Larva	Prepupa	Pupa	Male	Female
1.	20	1	6	—	9
2.	21	1	6	—	10
3.	19	2	6	6	—
4.	18	1	6	7	—
5.	21	1	7	—	11
6.	18	2	6	—	10
7.	21	1	7	—	10
8.	18	2	6	—	9
9.	21	1	6	—	10
10.	18	1	6	6	—
11.	18	1	6	—	11
12.	18	1	6	—	10
13.	18	1	6	6	—
14.	20	1	7	—	10
15.	18	1	6	6	—
16.	18	2	6	7	—
17.	18	1	6	—	12
18.	21	1	6	6.5	—
19.	19	1	6	—	10
20.	21	1	7	—	10
Mean	19.3 ± 1.36	1.2 ± 0.41	6.2 ± 0.41	6.35 ± 0.62	10.15 ± 0.78
Range	18 to 21	1 to 2	6 to 7	6 to 7	9 to 12

Third Instar

Larva was brownish black in colour with lateral hairs. The head capsule in 20 individuals measured 1.1 mm to 1.41 mm (mean 1.2 mm ± 0.094 mm) in width. Larvae measured from 13.2 mm to 14.00 mm (mean 13.7 mm ± 0.23 mm) in length and 4.00 mm to 4.4 mm (mean 4.2 mm ± 0.11 mm) in body width (Table 28). The instar lasted for 3 to 5 days (mean 4.2 ± 0.62 days) (Table 26).

Fourth Instar

The larva was elongate, colour became dark brown. The head capsules measured from 1.7 mm to 1.9 mm (mean 1.8 mm ± 0.73mm) in width. The larvae were measured from 22 mm to 24 mm (mean 23 mm ± 0.76 mm) in length and 4.3 mm to 4.7 mm (mean 4.5 mm ± 0.13 mm) in width (Table 28). The instar lasted for 4 to 5 days (mean 4.2 ± 0.41days) (Table 26).

Table 28: Morphometrics of different Stages of *H. produta*

Sl.No.	Larval									
	Head Width (mm)					Body Width (mm)				
	I	II	III	IV	V	I	II	III	IV	V
1.	0.45	0.95	1.1	1.8	2.3	0.92	2.3	4.0	4.7	5.0
2.	0.43	0.96	1.1	1.7	2.3	0.92	2.5	4.1	4.7	5.4
3.	0.44	0.95	1.2	1.9	2.3	0.95	2.4	4.3	4.5	5.7
4.	0.46	0.95	1.3	1.8	2.3	0.93	2.3	4.3	4.6	5.5
5.	0.45	0.93	1.1	1.7	2.4	0.93	2.9	4.4	4.5	5.7
6.	0.49	0.96	1.1	1.7	2.4	0.95	2.5	4.3	4.3	5.3
7.	0.45	0.95	1.1	1.9	2.4	0.95	2.4	4.3	4.5	5.7
8.	0.45	0.95	1.1	1.9	2.4	0.92	2.3	4.3	4.3	5.5
9.	0.43	0.93	1.2	1.8	2.3	0.92	2.2	4.1	4.3	5.4
10.	0.46	0.95	1.3	1.9	2.3	0.95	2.5	4.1	4.4	5.6
11.	0.45	0.97	1.3	1.7	2.5	0.94	2.4	4.1	4.3	5.6
12.	0.45	0.94	1.2	1.9	2.5	0.93	2.4	4.2	4.5	5.6
13.	0.47	0.95	1.3	1.8	2.4	0.92	2.3	4.1	4.6	5.6
14.	0.43	0.94	1.3	1.8	2.5	0.92	2.5	4.2	4.5	5.5
15.	0.44	0.95	1.2	1.8	2.4	0.93	2.4	4.2	4.6	5.5
16.	0.47	0.97	1.3	1.8	2.5	0.92	2.4	4.4	4.6	5.5
17.	0.43	0.94	1.1	1.9	2.4	0.93	2.5	4.2	4.6	5.6
18.	0.45	0.95	1.3	1.9	2.5	0.92	2.4	4.2	4.6	5.5
19.	0.45	0.95	1.4	1.8	2.4	0.93	2.5	4.2	4.6	5.5
20.	0.45	0.96	1.2	1.8	2.5	0.92	2.4	4.2	4.6	5.5
Mean	0.45±0.016	0.95±0.010	1.2±0.094	1.8±0.073	2.4±0.077	0.93±0.012	2.4±0.14	4.2±0.11	4.5±0.13	5.5±0.16
Range	0.43 to 0.47	0.90 to 0.93	1.1 to 1.4	1.7 to 1.9	2.3to 2.5	0.92 to 0.95	2.3 to 2.5	4.0 to 4.4	4.3 to 4.7	5.0 to 5.7

Table 28 (Continued): Morphometrics of different Stages of *H. produta*

Sl.No.	Instars Body Length (mm)					Prepupa		Pupa	
	I	II	III	IV	V	Width (mm)	Length (mm)	Width (mm)	Length (mm)
1.	3.0	5.0	14.0	22	28	5.0	30	4.7	11.0
2.	3.1	5.1	13.9	24	32	5.1	31	4.6	13.0
3.	3.0	5.2	13.9	23	32	5.1	30	4.6	12.0
4.	3.0	5.2	13.2	23	32	5.2	32	4.5	12.0
5.	3.1	5.1	13.0	22	33	5.2	30	4.6	13.0
6.	3.0	5.2	13.8	23	32	5.2	31	4.5	12.0
7.	2.7	5.3	13.8	22	31	5.0	30	4.6	11.0
8.	2.9	5.2	13.8	23	30	5.0	30	4.7	12.0
9.	3.0	5.2	13.7	23	31	5.0	32	4.6	13.0
10.	2.9	5.2	13.7	23	30	5.2	32	4.5	11.0
11.	2.7	5.3	13.8	23	32	5.0	31	4.7	13.0
12.	2.8	5.0	13.7	23	31	5.2	32	4.5	11.0
13.	2.8	5.3	13.8	24	31	5.2	31	4.6	13.0
14.	2.7	5.2	13.7	23	31	5.2	31	4.7	11.0
15.	3.0	5.3	13.8	24	32	5.0	32	4.5	13.0
16.	3.0	5.3	13.8	23	30	5.1	31	4.6	12.0
17.	3.0	5.1	13.8	24	30	5.1	32	4.7	11.0
18.	2.8	5.3	13.7	22	31	5.1	30	4.5	12.0
19.	2.9	5.3	13.7	24	30	5.1	32	4.7	12.0
20.	2.8	5.2	13.5	22	31	5.0	30	4.6	12.0
Mean	2.9± 0.13	5.2± 0.097	13.7 ± 0.23	23± 0.76	31 ± 1.02	5.1± 0.26	31 ± 1.1	4.6± 0.08	12± 0.83
Range	2.7 to 3.1	5.0 to 5.3	13.2 to 14	22 to 24	30 to 33	5.0 to 5.2	30 to32	4.5 to 4.7	11 to 13

Table 28 (Continued): Morphometrics of different Stages of *H. produta*

Sl.No.	Male			Adult			Female	
	Body Length (mm)	Body Width (mm)	Wing Expansion (mm)	Body Length (mm)	Body Width (mm)	Wing Expansion (mm)		
1.	21	3.8	50	22	3.8	54		
2.	20	3.6	52	21	4.0	54		
3.	20	3.8	52	21	4.0	54		
4.	20	3.7	50	22	3.9	53		
5.	20	3.8	51	23	3.9	53		
6.	21	3.7	51	22	3.8	54		
7.	21	3.6	50	23	4.0	55		
8.	20	3.6	52	22	3.9	55		
9.	20	3.8	52	23	3.8	55		
10.	21	3.7	52	23	4.0	53		
11.	20	3.6	52	22	3.8	54		
12.	21	3.8	50	21	3.9	55		
13.	20	3.6	51	22	3.9	53		
14.	21	3.7	51	21	3.8	54		
15.	20	3.8	50	23	4.0	53		
16.	21	3.6	51	22	4.0	54		
17.	21	3.8	51	21	4.0	55		
18.	21	3.6	50	22	3.8	55		
19.	20	3.7	51	21	3.9	53		
20.	21	3.7	51	23	3.8	55		
Mean	20.5 ± 0.51	3.7 ± 0.086	51 ± 0.83	22 ± 0.83	3.9 ± 0.86	54 ± 0.83		
Range	20 to 21	3.6 to 3.8	50 to 52	21 to 23	3.8 to 4.0	53 to 55		

Fifth Instar

The larva became elongated and body surface showed very long hairs. Head was black. Prolegs were conspicuous. Head capsule measured from 2.3mm to 2.5 mm (mean 2.4 mm ± 0.077 mm) in width (Table 28). Larvae ranged from 30 mm to 33 mm (mean 31 mm ± 1.02 mm) in body length and 5 mm to 5.7 mm (mean 5.5 mm ± 0.16 mm) in body width. This instar lasted for 4 to 5 days (mean 4.8 ± 0.41 days) (Table 26).

Instars of *H.producta* showed increased head capsule, body width and body length. The instars and their size was tested by applying regression analysis and found positively correlated and represented in Tables 29–31 and Figures 137–139.

Table 29: Statistics of Linear Regression Relationship of Larval Instars and Larval Head Width of *H. producta*

Sl.No.	Instar X	X²	Head Width Y	Y²	XY	Expected Values Y
1.	1	1	0.45	0.2025	0.45	0.41
2.	2	4	0.95	0.9025	1.9	0.88
3.	3	9	1.2	1.44	3.6	1.36
4.	4	16	1.8	3.25	7.2	1.835
5.	5	25	2.4	5.76	12.00	2.31
Σ	3		1.36			

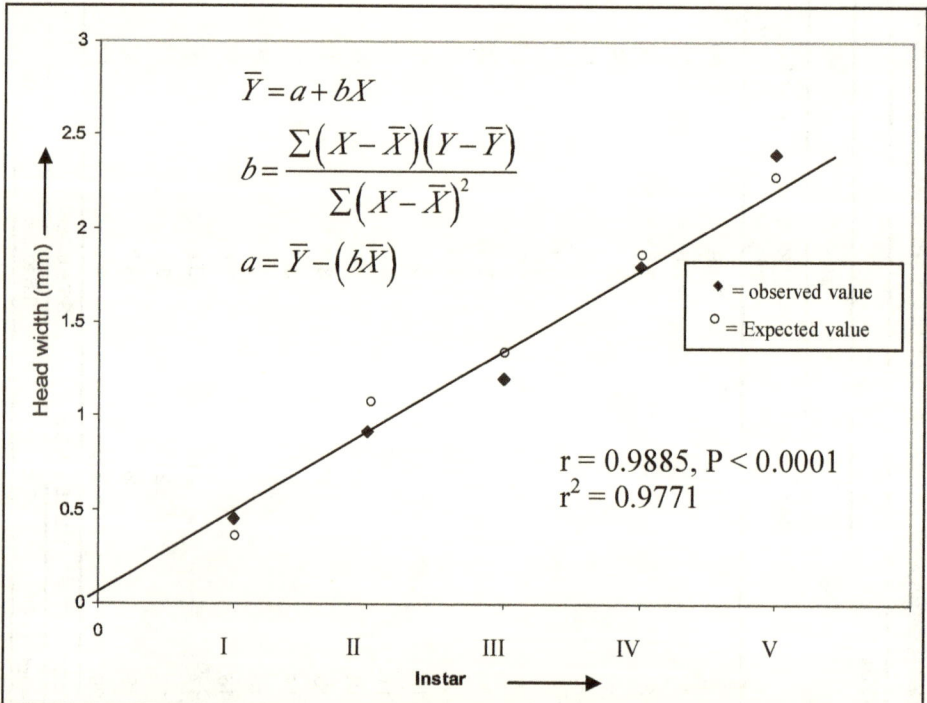

$$\overline{Y} = a + bX$$

$$b = \frac{\Sigma\left(X - \overline{X}\right)\left(Y - \overline{Y}\right)}{\Sigma\left(X - \overline{X}\right)^2}$$

$$a = \overline{Y} - \left(b\overline{X}\right)$$

♦ = observed value
○ = Expected value

$r = 0.9885, P < 0.0001$
$r^2 = 0.9771$

Figure 137: Relation between Larval Head Width and Larval Instars of *H. produta*.

Table 30: Statistics of Linear Regression Relationship of Larval Instars and Larval Body Width of *H. producta*

Sl.No.	Instar X	X^2	Body Width Y	Y^2	XY	Expected Values Y
1.	1	1	0.93	0.8649	0.93	1.259
2.	2	4	2.4	5.76	4.8	2.113
3.	3	9	4.2	17.64	12.6	3.507
4.	4	16	4.5	20.25	18.0	4.631
5.	5	25	5.5	30.25	27.5	5.755
Σ	3		3.506			

Table 31: Statistics of Linear Regression Relationship of Larval Instars and Larval Body Length of *H. producta*

Sl.No.	Instar X	X^2	Body Length Y	Y^2	XY	Expected Values Y
1.	1	1	2.9	8.41	2.9	0.36
2.	2	4	5.2	27.04	10.4	7.76
3.	3	9	13.7	187.69	41.1	15.16
4.	4	16	23.0	529.00	92.00	22.56
5.	5	25	31.0	961.00	155.00	29.96
Σ	3		15.16			

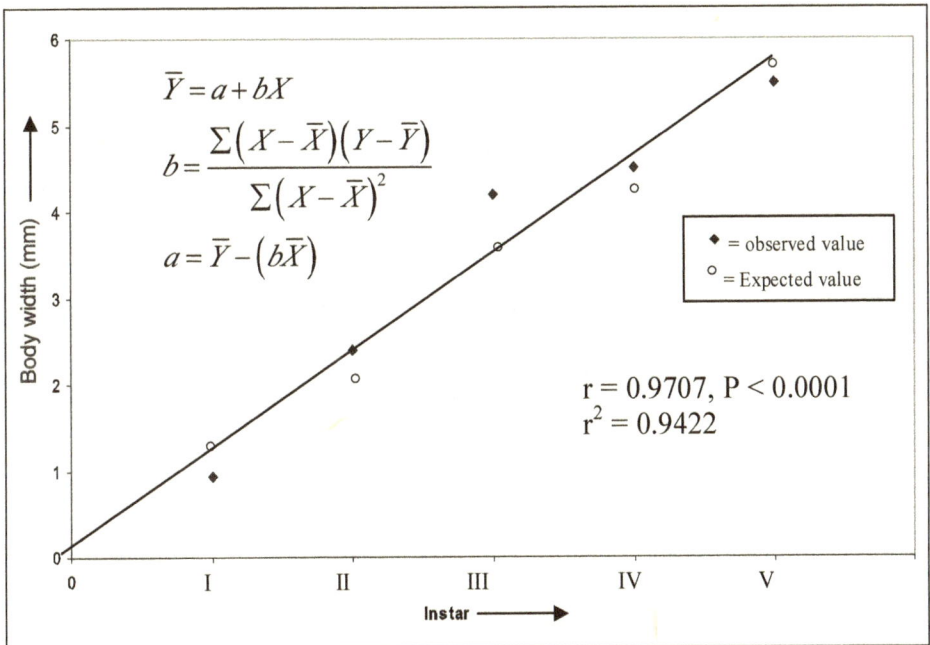

$$\bar{Y} = a + bX$$

$$b = \frac{\sum(X - \bar{X})(Y - \bar{Y})}{\sum(X - \bar{X})^2}$$

$$a = \bar{Y} - (b\bar{X})$$

♦ = observed value
○ = Expected value

r = 0.9707, P < 0.0001
$r^2 = 0.9422$

Figure 138: Relation between Larval Body Width and Larval Instars of *H. produta*.

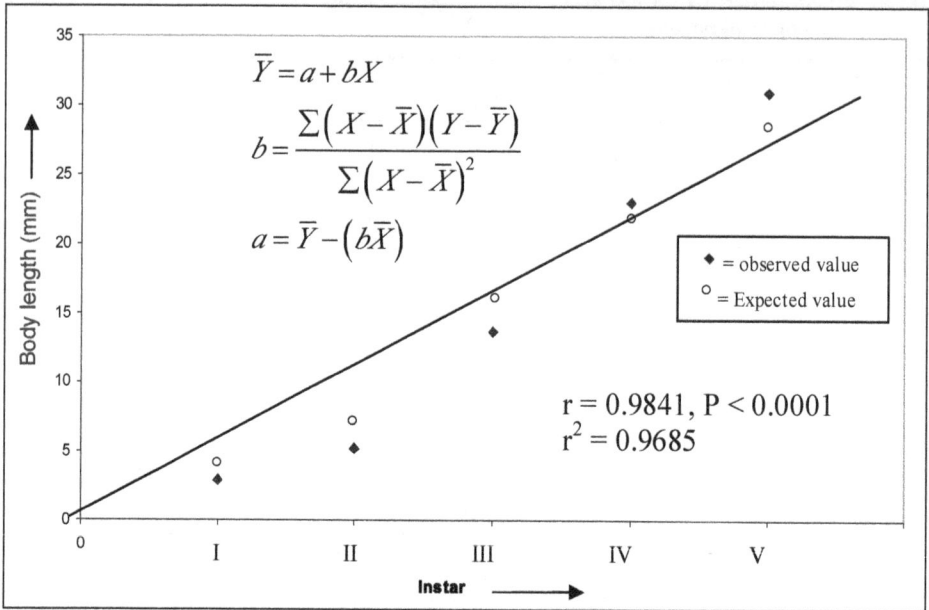

$$\bar{Y} = a + bX$$

$$b = \frac{\Sigma(X - \bar{X})(Y - \bar{Y})}{\Sigma(X - \bar{X})^2}$$

$$a = \bar{Y} - (b\bar{X})$$

♦ = observed value
○ = Expected value

$r = 0.9841, P < 0.0001$
$r^2 = 0.9685$

Figure 139: Relation between Larval Body Length and Larval Instars of *H. produta.*

Nature of Damage and Larval Behaviour

Newly hatched larva fed on tender leaves of *F.glomerata*. Second instar larva fed voraciously on tissues between the network of veins. Later instars migrated from one leaf to another and fed on green matter of tender leaves. During heavy infesttion 2 to 3 larvae can be seen on the rachis of leaf.

Prepupa

The full grown larva was sluggish and stopped its feeding. The larva gradually shortened to the size of future pupa. Prepupal period lasted for 1 to 2 days (mean 1.2 ± 0.41 days) (Table 27).

Pupa

The larva pupated on either lower or upper surface of leaf. Freshly formed pupa was dark reddish brown in colour. In 20 individuals pupal length ranged from 11 mm to 12 mm (mean 12 mm ± 0.83mm) and width ranged from 4.5 mm to 4.7 mm (mean 4.6 mm ± 0.08 mm) (Table 28). Pupal stage lasted for 6 to 7 days (mean 6.2 ± 0.41 days) (Table 27).

Female Adult

Female moth was medium sized and yellow in colour. The abdomen was bulging and measured from 21 mm to 23 mm (mean 22 ± 0.83 mm) in length and 3.8 mm to 4 mm (mean 3.9 mm ± 0.86 mm) in width with black spots. Wing expansion ranged from 53 mm to 55 mm (mean 54 mm ± 0.83 mm) (Table 28). Adult longevity varied from 9 to 12 days (mean 10.15 ± 0.97 days) (Table 27).

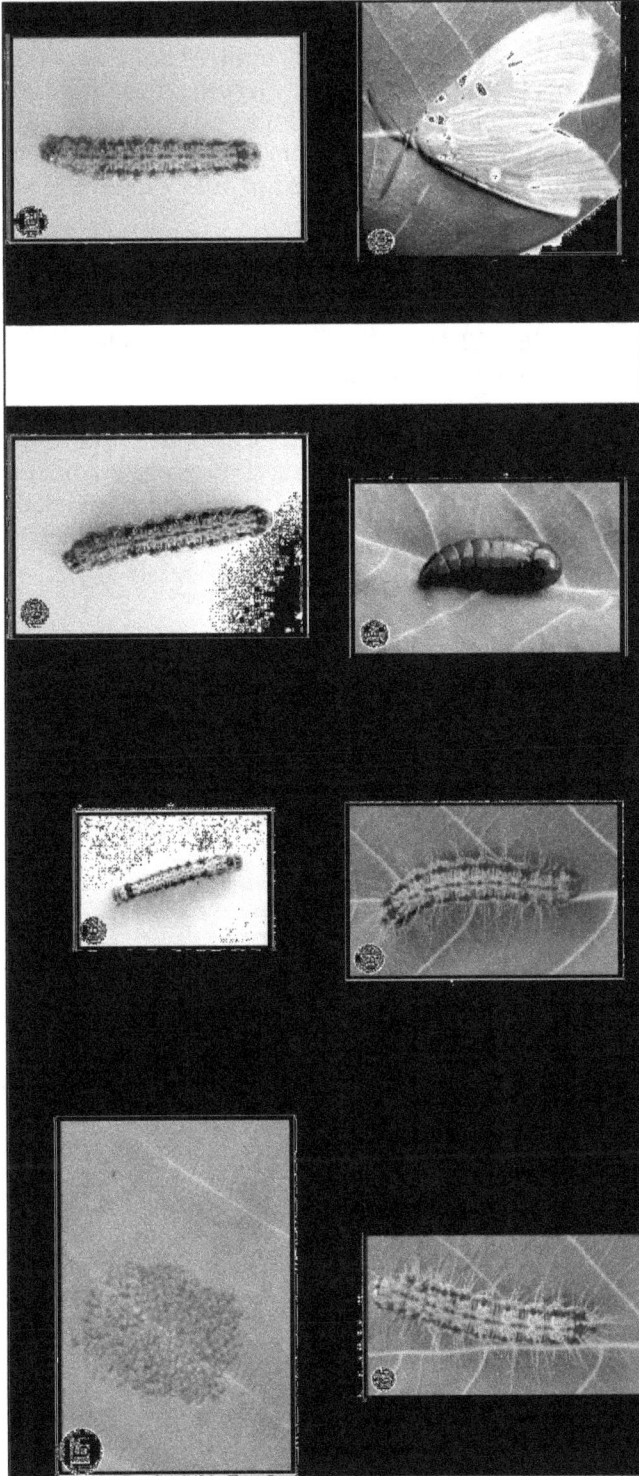

PLATE 39—Life Cycle of *Hypsa producta.* Figure 161: Eggs; Figure 162: First instar larva; Figure 163: Second instar larva; Figure 164: Third instar larva; Figure 165: Fourth instar larva; Figure 166: Fifth instar larva; Figure 167: Pupa; Figure 168: Adult.

Male Adult

Male moth was smaller than female, stoutly built. Abdomen was slender and pointed towards anal end. In 20 individuals body length ranged from 20 mm to 21 mm (mean 20.5 mm ± 0.51 mm) while the width ranged from 3.6 mm to 3.8 mm (mean 3.7 mm ± 0.086 mm). Wing expansion ranged from 50 mm to 52 mm (mean 51 mm ± 0.83 mm) (Table 28). Adult longevity varied from 6 to 7 days (mean 6.35 ± 0.62 days) (Table 27).

Life Cycle

Duration of egg, larval, prepupal, pupal and adult stages varied from 3 to 3.5 days (mean 3 ± 0.23 days), 16 to 22 days (mean 19.1 ± 1.36 days), 1 to 2 days (mean 1.2 ± 0.41 days), 6 to 7 days (mean 6.2 ± 0.41 days) and 9 to 12 days (mean 10.15 ± 0.97 days) respectively. Larvae completed its development in 19.1 ± 1.36 days by passing through five instars. The duration of which lasted for 2.3 ± 0.44, 3.6 ± 0.50, 4.2 ± 0.62, 4.2 ± 0.41 and 4.8 ± 0.41 days respectively. The life cycle from egg to adult varied from 26 to 34.5 days (mean 29.50 ± 2.41 days). Period required for completion of one generation varied from 35 to 46.5 days (mean 39.65 ± 3.38 days) (Table 32).

Sex Ratio

Sex ratio, male to female averaged 1:2.03 (Table 33).

Table 32: Life Cycle of *H. produta*

Stages	Duration in Days		
	Min.	Max.	Mean
Egg	3	3.5	3.0 ± 0.23
Larval Instars			
First instar	2	3	2.3 ± 0.44
Second instar	3	4	3.6 ± 0.50
Third instar	3	5	4.2 ± 0.62
Fourth instar	4	5	4.2 ± 0.41
Fifth instar	4	5	4.8 ± 0.41
Total larval period	16	22	19.1 ± 1.36
Pre Pupal period	1	2	1.2 ± 0.41
Pupal period	6	7	6.2 ± 0.41
Adult Male	6	7	6.35 ± 0.62
Adult Female	9	12	10.15 ± 0.97
Life cycle	26	34.5	29.5 ± 2.41
Period of generation	35	46.5	39.65 ± 3.38

Table 33: Sex Ratio of *H. produta*

Sl.No.	Adult Examined	Males	Females
1.	10	5	5
2.	10	4	6
3.	10	3	7
4.	10	2	8
5.	10	4	6
6.	10	5	5
7.	10	3	7
8.	10	2	8
9.	10	2	8
10.	10	3	7
Total	100	33	67

Sex ratio for male to female = 1:2.03

Intrinsic Rates of Increase

1. Intrinsic Rates of Increase in *E. narcissus*

The first adult mortality was noted on the 8th day. Average period of immature stages was 33 days, Maximum mean progeny production per day, m x was 20.5 on the 3rd day. The innate capacity for increase was found to be 0.129 (Figure 140) per female per day and population of *E. narcissus* multiplied 75.78 times in generation time 'T' of 33.54 days. Results are shown in Tables 34–38.

$$T_c = \frac{1_x m_x X}{1_x m_x} = \frac{2766.98}{75.78} = 36.51$$

where, T_c is arbitrary T.

$$\frac{= \log_e R_0}{= T_c} = \frac{75.78}{36.51} = 0.118$$

where, r_c is arbitrary r_m

$T_c = 36.51$

$R_c = 0.118$

Now arbitrary 'r_m's are 0.09 and 0.013 where λ is the finite rate of natural increase.

$$T = \frac{\log_e 75.78}{0.129} = 33.54$$

$T = 33.54$ *days*

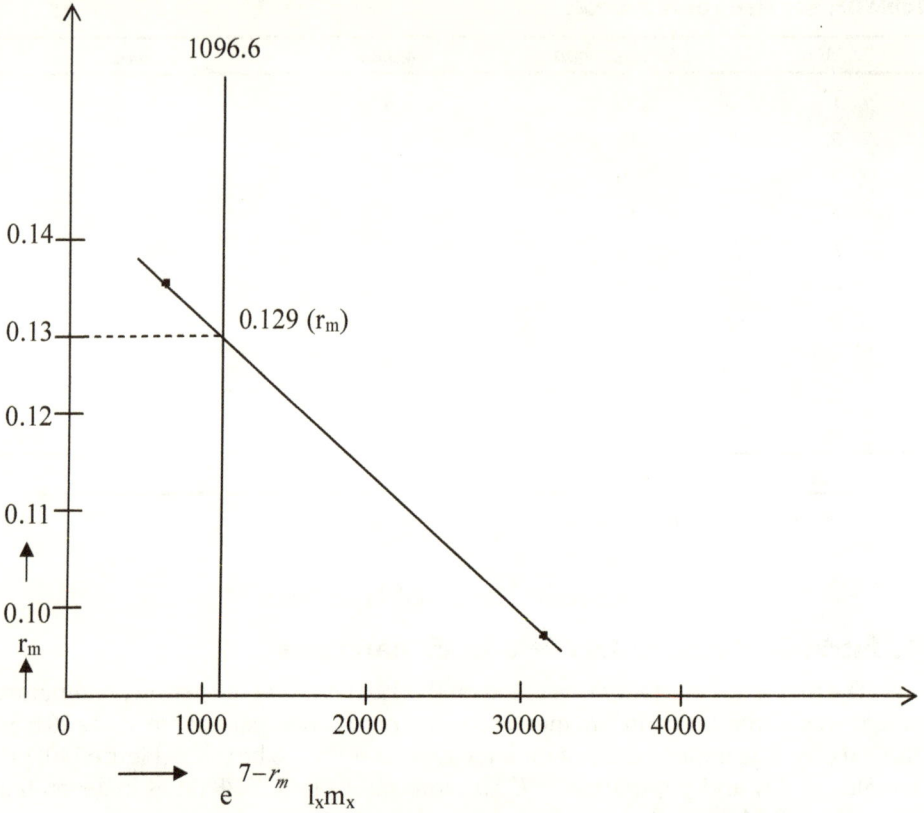

Figure 140: Determination of Intrinsic Rate of Increase in *E. narscissus.*

Table 34: Developmental Period Required for Female of *E. narcissus*

Sl.No.	Egg	Larva	Pupa	Adult Formation (Total Days)
1	4	21	6	31
2	4	24	6	34
3	3	23	7	33
4	3	21	7	31
5	4	21	8	33
6	3	22	7	32
7	3	23	6	32
8	3	23	8	34
9	4	22	6	32
10	3	23	9	35
Mean				32.7

Table 35: Daily Production of Progeny by Mated Females of *E. narcissus*

Female Number	Replicates																		Males	Fe-males	Total
	Number of Progeny Produced/Day																				
	1		2		3		4		5		6		7		8		9				
	M	F	M	F	M	F	M	F	M	F	M	F	M	F	M	F	M	F			
1	6	7	5	16	10	15	9	9	4	10	D	7	–	5	–	D	–	–	34	69	103
2	6	9	7	12	12	19	8	14	7	10	D	6	–	7	–	D	–	–	40	77	117
3	7	7	6	19	13	20	6	15	5	10	3	7	D	3	–	D	–	–	40	81	121
4	6	5	5	9	9	21	10	17	4	10	D	5	–	3	–	D	–	–	34	70	104
5	6	11	5	21	11	22	9	9	7	8	D	5	–	3	–	D	–	–	38	79	117
6	5	7	4	14	10	22	8	12	5	10	3	6	–	4	–	D	–	–	35	75	110
7	4	8	6	16	9	20	10	11	6	8	D	8	–	5	–	D	–	–	35	76	111
8	5	5	7	12	11	21	9	12	5	8	D	8	–	7	–	5	–	D	37	78	115
9	5	8	8	16	9	25	9	10	6	7	D	6	–	5	–	D	–	–	37	77	114
10	5	6	5	19	11	20	12	12	6	8	D	8	–	6	–	4	–	D	39	83	122
Mean	5.5	7.3	5.8	15.4	10.5	20.5	9.0	12.1	5.5	8.9	0.6	6.6	–	4.8	–	0.9	–	–	36.9	76.5	113.4

Table 36: Life Table Statistics of *E. narcissus*

Pivotal Age (Days) x	Propotional Live Age x lx	Number of Female Progeny per Female m_x	$l_x m_x$	$l_x m_x x$
1-33 days immature stages				
34	1.0	7.30	7.30	248.2
35	1.0	15.40	15.40	539.0
36	1.0	20.50	20.50	738.0
37	1.0	12.1	12.1	447.7
38	1.0	8.9	8.9	338.2
39	1.0	6.6	6.6	257.4
40	1.0	4.8	4.8	192.0
41	0.2	0.9	0.18	7.38
42	0.0	0.0	0.0	0.00
43	0.0	0.0	0.0	0.00
			Σ 75.78	Σ 2766.98

Table 37: Provisional r_m (0.09) for *E. narscissus* and Related Values of $e^{7-r} mx l_x m_x$

x	r_{mx}	$e^{7-r} mx$	$e^{7-r} mx$	$e^{7-r} mx l_x m_x$
34	3.06	3.94	51.41	375.29
35	3.15	3.85	46.99	723.64
36	3.24	3.76	42.94	880.27
37	3.33	3.67	39.25	474.92
38	3.42	3.58	35.87	319.24
39	3.51	3.49	32.78	216.34
40	3.60	3.40	29.96	143.80
41	3.69	3.31	27.38	4.92
42	3.78	3.22	25.02	0.0
43	3.87	3.13	22.87	0.0
				Σ3138.42

Table 38: Provisional r_m (0.13) for *E. narscissus* and Related Values of $e^{7-r} mx l_x m_x$

x	r_{mx}	$e^{7-r} mx$	$e^{7-r} mx$	$e^{7-r} mx l_x m_x$
34	4.42	2.58	13.19	96.28
35	4.55	2.45	11.58	178.31
36	4.68	2.32	10.17	208.48
37	4.81	2.19	8.93	108.05

Contd...

Table 38–Contd...

x	r_{mx}	$e^{7-r}mx$	$e^{7-r}\,mx$	$e^{7-r}\,mx/\,m_x$
38	4.94	2.06	7.84	69.77
39	5.07	1.93	6.88	45.40
40	5.2	1.8	6.04	28.99
41	5.33	1.67	5.31	0.95
42	5.46	1.54	4.66	0.0
43	5.59	1.41	4.09	0.0
				$\Sigma 736.23$

2. Intrisic Rates of Increase in *E. machaeralis*

The first adult mortality was noted on the 6th day. Average period of immature stages was 28 days, Maximum mean progeny production per day, m x was 26.8 on the 3rd day. The innate capacity for increase was found to be 0.149 (Figure 141) per female per day and population of *E. machaeralis* multiplied 88.16 times in generation time 'T' of 30.6 days. Results are shown in Tables 39–43.

where, T_c is arbitrary T.

$$T_c = \frac{1_x m_x \, X}{1_x m_x} = \frac{2743.2}{88.16} = 31.11$$

where, r_c is arbitrary r_m

$$= \frac{\log_e R_0}{T_c} = \frac{88.16}{31.11} = 0.143$$

$T_c = 31.11$

$R_c = 0.143$

Now arbitrary 'r_m's are 0.12 and 0.16 where λ is the finite rate of natural increase.

$r_m = 0.149$ (figure)

$$T = \frac{\log_e 88.16}{0.149} = 30.06$$

$T = 30.06$ days.

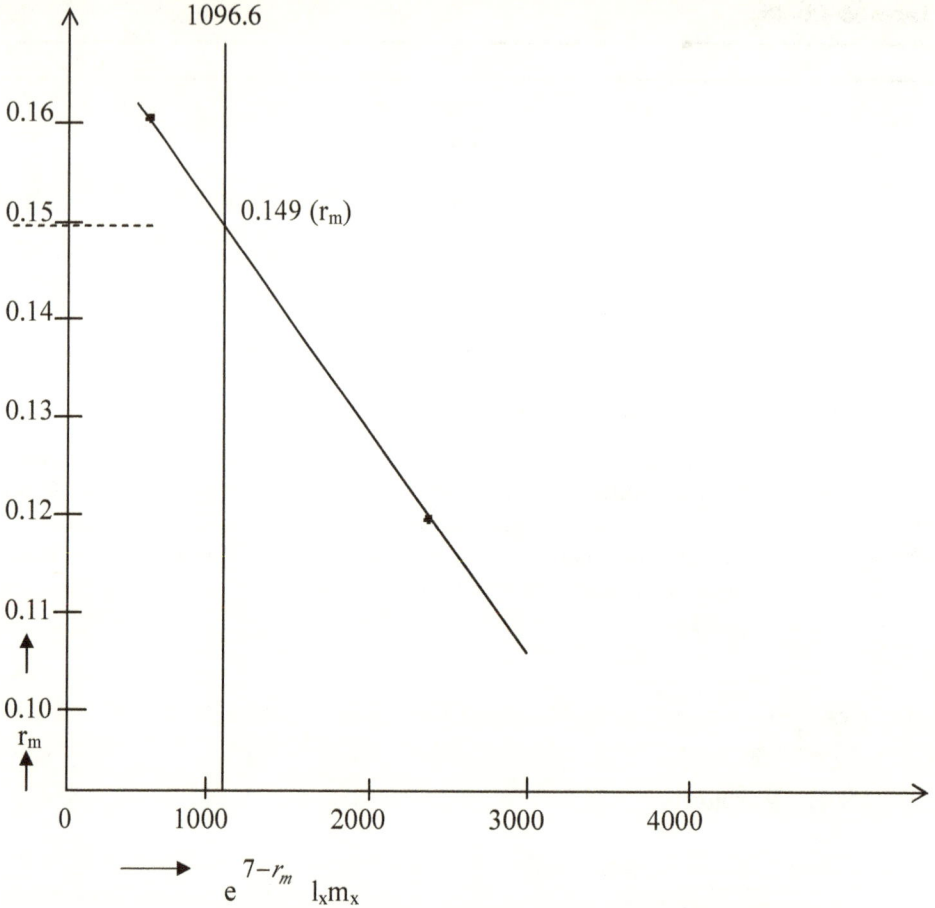

Figure 141: Determination of Intrinsic Rate of Increase in *E. machaeralis.*

Table 39: Developmental Period Required for Females of *E. machaeralis*

Sl.No.	Egg	Larva	Pupa	Adult Formation (Total Days)
1.	3	18	8	29
2.	3	18	6	27
3.	3	16	7	26
4.	3	18	6	27
5.	4	17	7	28
6.	3	18	8	29
7.	3	16	7	26
8.	3	17	8	28
9.	3	18	8	29
10.	3	18	7	28
Mean				27.7

Table 40: Daily Production of Progeny by Mated Females of *E. machaeralis*

Replicates																	Number of Progeny Produced/Day			
Female Number	1		2		3		4		5		6		7		8		Males	Fe-males	Total
	M	F	M	F	M	F	M	F	M	F	M	F	M	F	M	F			
1	8	12	9	19	21	31	11	14	3	5	D	D	–	5	–	–	52	81	133
2	5	11	11	13	21	25	9	15	6	11	5	5	D	5	–	D	57	85	142
3	7	14	12	16	19	33	13	14	3	4	D	D	–	5	–	–	54	81	135
4	6	11	11	19	20	21	14	18	8	13	3	9	D	5	–	D	62	96	158
5	7	13	12	23	18	22	14	17	5	9	2	3	D	D	–	–	58	87	145
6	5	10	14	19	23	27	11	19	8	13	3	8	D	D	–	–	64	96	160
7	8	15	12	18	19	26	11	14	7	12	5	6	D	D	–	–	60	91	151
8	9	10	13	19	21	28	9	15	5	14	D	D	–	–	–	–	57	86	143
9	7	12	12	18	23	25	14	14	4	12	5	10	D	6	–	D	65	97	162
10	10	12	13	19	21	30	9	16	6	11	D	D	D	–	–	–	59	88	147
Mean	7.2	12	11.9	18.3	20.6	26.8	11.5	15.6	5.5	10.4	2.3	4.1	–	1.6	–	–	58.8	88.8	1476

Table 41: Life Table Statistics of *E. machaeralis*

Pivotal Age (Days) x	Propotional Live Age x lx	Number of Female Progeny per Female m_x	$l_x m_x$	$l_x m_x x$
1-28 days immature stages				
29	1.0	12.0	12.00	348.00
30	1.0	18.30	18.30	549.00
31	1.0	26.80	26.80	830.80
32	1.0	15.60	15.60	499.20
33	1.0	10.40	10.40	343.20
34	1.0	4.10	4.10	139.40
35	0.6	1.60	0.96	33.60
36	0.9	0.0	0.00	0.00
37	0.0	0.0	0.00	0.00
			Σ88.16	Σ2743.2

Table 42: Provisional r_m (0.12) for *E. machaeralis* and Related Values of $e^{7-r} mxl_x m_x$

x	r_{mx}	$e^{7-r}mx$	$e^{7-r} mx$	$e^{7-r} mxl_x m_x$
29	3.48	3.52	33.78	405.36
30	3.6	3.4	29.96	548.268
31	3.72	3.28	26.57	712.076
32	3.84	3.16	23.57	367.602
33	3.96	3.04	20.90	217.36
34	4.08	2.92	18.54	76.014
35	4.2	2.8	16.44	15.78
36	4.32	2.68	14.58	0
37	4.44	2.56	12.93	0
				Σ2342.46

Table 43: Provisional r_m (0.16) for *E. machaeralis* and Related Values of $e^{7-r} mxl_x m_x$

x	r_{mx}	$e^{7-r}mx$	$e^{7-r} mx$	$e^{7-r} mxl_x m_x$
29	4.64	2.36	10.59	127.08
30	4.8	2.2	9.02	165.06
31	4.96	2.04	7.69	206.09
32	5.12	1.88	6.55	102.18
33	5.28	1.72	5.58	58.03
34	5.44	1.56	4.75	19.47
35	5.6	1.4	4.05	3.88
36	5.76	1.24	3.45	0.0
37	5.92	1.08	2.94	0.0
				Σ 681.79

3. Intrisic Rates of Increase in *H. producta*

The first adult mortality was noted on the 5th day. Average period of immature stages was 30 days, Maximum mean progeny production per day, M_x was 26 on the 3rd day. The innate capacity for increase was found to be 0.141 (Figure 142) per female per day and population of *H. producta* multiplied 76.76 times in generation time 'T' of 30.78 days. Results are shown in Tables 44–48.

$$T_c = \frac{1_x m_x X}{1_x m_x} = \frac{2506.34}{76.76} = 32.65$$

where, T_c is arbitrary T.

$$= \frac{\log_e R_0}{T_c} = \frac{76.76}{32.65} = 0.132$$

where, r_c is arbitrary r_m

$T_c = 32.65$

$r_c = 0.132$

Now arbitrary 'r_m's are 0.11 and 0.15 where λ is the finite rate of natural increase.

$r_m = 0.141$ (figure)

$$T = \frac{\log_e 76.76}{0.141} = 30.78$$

$T = 30.78$ days.

Table 44: Developmental Period Required for Females of *H. producta*

Sl.No.	Egg	Larva	Pupa	Adult Formation (Total Days)
1.	3	20	7	30
2.	3	21	7	31
3..	3	21	8	32
4.	3	18	8	29
5.	3	21	8	32
6.	3	18	7	28
7.	3	21	7	31
8.	3	18	7	28
9.	3	18	7	28
10.	3	20	8	31
Mean				30

Table 45: Daily Production of Progeny by Mated Females of *H. producta*

Replicates															Number of Progeny Produced/Day				
Female Number	1		2		3		4		5		6		7		Males	Females	Total
	M	F	M	F	M	F	M	F	M	F	M	F	M	F			
1	7	12	8	18	17	28	5	14	2	–	D	7	–	D	39	79	118
2	9	14	10	16	12	19	5	17	2	10	D	D	–	–	38	76	114
3	8	17	9	21	15	25	5	17	3	D	D	5	–	–	40	80	120
4	6	12	10	20	15	24	5	10	5	11	D	D	–	D	41	82	123
5	9	14	11	17	14	23	4	15	D	6	–	D	–	D	38	75	113
6	8	11	10	18	12	27	4	13	3	4	–	D	–	–	37	73	110
7	7	15	12	27	16	28	7	15	D	D	–	D	–	–	42	85	127
8	8	13	12	19	12	29	6	14	3	8	D	D	–	D	41	83	124
9	9	12	11	27	12	30	6	9	D	D	–	D	–	–	38	78	116
10	10	16	11	19	12	27	6	11	D	7	–	D	–	–	39	80	119
Mean	8.1	13.6	10.4	20.2	13.7	26.0	5.3	13.5	1.8	4.6	–	1.2	–	–	39.3	79.1	118.4

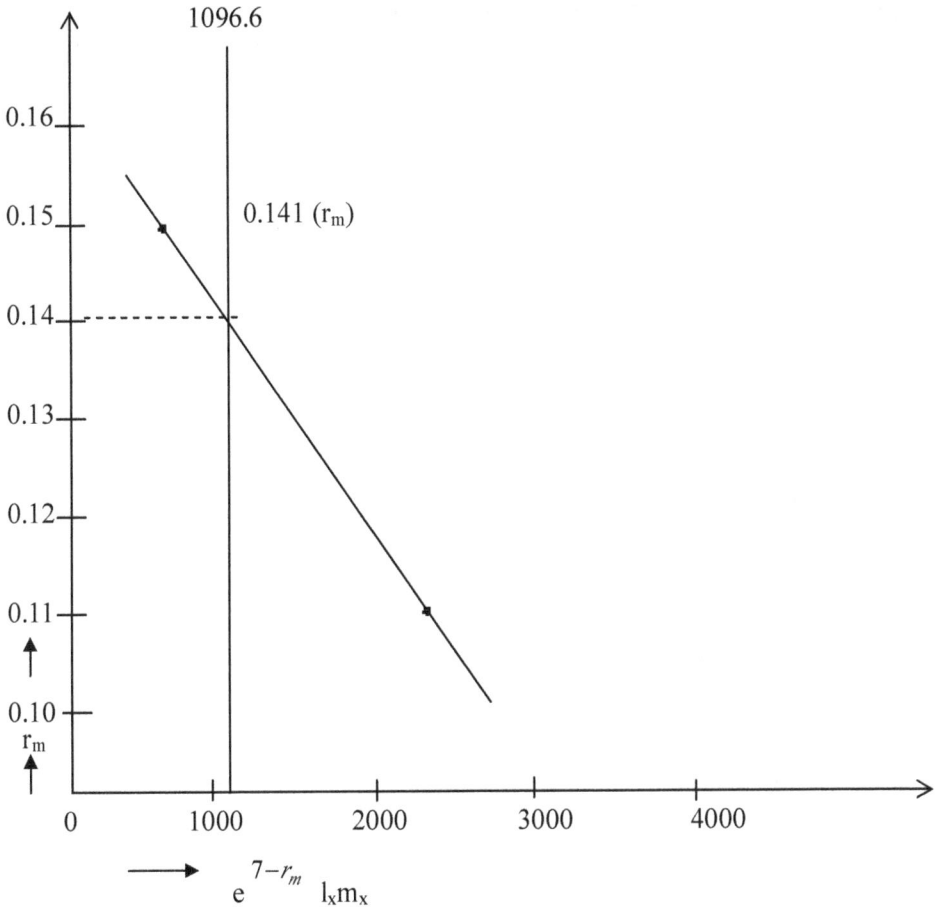

Figure 142: Determination of Intrinsic Rate of Increase in *H. producta*.

Table 46: Life Table Statistics of *H. producta*

Pivotal Age (Days) x	Propotional Live Age x lx	Number of Female Progeny per Female m_x	$l_x m_x$	$l_x m_x x$
1-30 days immature stages				
31	1.0	13.6	13.60	421.60
32	1.0	20.2	20.20	646.40
33	1.0	26.0	26.00	858.00
34	1.0	13.5	13.50	459.00
35	0.7	4.6	3.22	112.70
36	0.2	1.2	0.24	8.64
37	0.0	0.0	0.00	0.00
			Σ 76.76	Σ 2506.34

Table 47: Provisional r_m (0.11) for *H. producta* and Related Values of $e^{7-r} mxl_xm_x$

x	r_{mx}	$e^{7-r} mx$	$e^{7-r}\ mx$	$e^{7-r}\ mxl_xm_x$
31	3.41	3.59	36.23	492.72
32	3.52	3.48	32.45	655.49
33	3.63	3.37	29.07	755.82
34	3.74	3.26	26.04	351.54
35	3.85	3.15	23.33	75.12
36	3.96	3.04	20.90	5.01
37	4.07	2.93	18.72	0.0
			Σ 2335.7	

Table 48: Provisional r_m (0.5) for *H. producta* and Related Values of $e^{7-r} mxl_xm_x$

x	r_{mx}	$e^{7-r} mx$	$e^{7-r}\ mx$	$e^{7-r}\ mxl_xm_x$
31	4.65	2.33	10.48	142.52
32	4.8	2.20	9.02	182.20
33	4.95	2.05	7.76	201.76
34	5.1	1.90	6.68	90.18
35	5.25	1.75	5.75	18.51
36	5.4	1.60	4.95	1.18
37	5.55	1.45	4.26	0.0
			Σ 636.35	

6

Checklist of Moths

Checklist has several objectives such as scrutiny, inspection, monitoring, investigation, enquiry, thing confirmation, verification etc. The present topic is aimed for monitoring and investigating the moth species in the western Ghats of Kolhapur and Satara district of Maharashtra, India. The present work will be helpful for understanding the biodiversity of moths in the districts and counting the status of the moths.

Checklist of moths from western Ghats of Kolhapur and Satara district has been given in Table 49.

Table 49: Checklist of Moths from Western Ghats of Kolhapur and Satara Districts (Maharashtra)

Sl.No.	Moth Species		Host Plants	
	Common Name	*Generic Name*	*Common Name*	*Generic Name*
Family – Arctiid ae				
1.	Arctiid moth	*Argina guttata* Hampson	Sweet lime	*Citrus limettia Risso*
2.	Arctiid moth	*Argina kolhapurensis* sp. nov.	Fig	*Ficus religiosa*
3.	Arctiid moth	*Argina indica* sp. nov.	Fig	*F. religiosa*
4.	Arctiid moth	*Argina satarensis* sp. nov.	Tamarind (Chinch)	*Tamarindus indica* L.
5.	Arctiid moth	*Spilosoma sathei* sp. nov.	Common beans	*Phaseolus vulgaris* L.
6.	Arctiid moth	*Creatonotus gangis* L.	Pomegranate	*Punica grantum*
7.	Arctiid moth	*Asura ruptofascia* Hampson	Mango (Amba)	*Mangifera indica* L.
8.	Arctiid moth	*Pelochyta sathei* sp.nov.	Mango (Amba)	*M. indica*

Contd...

Table 49–Contd...

Sl.No.	Moth Species		Host Plants	
	Common Name	Generic Name	Common Name	Generic Name
9.	Arctiid moth	*Moorea marathi* sp.nov.	Wild Banana	*Musa acuminata* Colla.
10.	Arctiid moth	*M. indica* sp.nov.	Banana	*M. acuminata*
11.	Bihar hairy caterpillar	*S. obliqua* Walker	White Mulberry	*Morus alba* L.
12.	Black wooly caterpillar	*Amsacta lactinea* Cramer	Teak	*Tectona grandis* Linn.
13.	Sunn hemp Hairy caterpillar	*Utetheisa pulchella* Linn.	*Sunn hemp*	*Crotalaria juncea* L.
14.	Arctiid moth	*Celama fasciata* Walker	Mango (Amba)	*M. indica*

Family – Hypsidae

15.	Hypsid moth	*Hypsa producta* Moore	Fig	*Ficus glomerata*
16.	Hypsid moth	*Hypsa marathi* sp. nov.	Fig	*F. glomerata*

Family – Gelechidae

17.	Gelechid moth	*Anarsia lineatella* Zeller	Mango (Amba)	*M. indica*
18.	Gelechid moth	*A. sagittaria* Meyrick	Ber	*Ziziphus jujuba* Mill
19.	Gelechid moth	*Chelaria* sp.	Sapota	*Achras sapota*
20.	Gelechid moth	*Holcocera pulverea* (Meyrick)	Tamarind (Chinch)	*T. indica*
21.	Gelechid moth	*Sitotroga cerealella* Olivier	Fig	*Ficus recemosa* L.
22.	Gelechid moth	*Hypatima haligramma*	Mango	*M. indica*

Family – Caprosinidae

23.	Caprosinid moth	*Meridarchis reprobata* Meyrick	Ber / Jamun	*Z. jujuba* / *Syzygium cuminii* L.
24.	Caprosinid moth	*Meridarchis scyrodes* Meyrick	Ber	*Z. jujuba*

Family – Sphingidae

25.	Sphingid moth	*Cephanodes indica* sp. nov.	Behada	*Terminalia bellirica* Roxb.
26.	Sphingid moth	*Acherontia styx* Westwood	Mango (Amba)	*M. indica*
27.	Sphingid moth	*Theretra pallicosta* (Walker)	Jangli avala	*Phylathus emblica*
28.	Sphingid moth	*Theretra nessus* Drury	Palash	*Butea monosperma* L.
29.	Sphingid moth	*T. pinastrina pinastrina* Mart	Saapkanda	*Arisoema sahyadricum* L.
30.	Sphingid moth	*T. aceta* Cramer	Saapkanda	*A. sahysdricum* L.
31.	Sphingid moth	*T. alecto*	Madder	*Rubia tinctorum* L.
32.	Sphingid moth	*Hippotion velox* Fab.	Catch bird Tree	*Pisonia aculeata*
33.	Sphingid moth	*Cornpsogene panopus* Cramer	Mango	*M. indica*

Contd...

Table 49—Contd...

Sl.No.	Moth Species		Host Plants	
	Common Name	Generic Name	Common Name	Generic Name
34.	Sphigid moth	*Hippotion rafftera* Butl.	Balsam	*Impatiens balsamina* Linn.
35.	Sphingid moth	*H. celerio* L.	Indian Almond	*Terminalia catappa* L.
36.	Sphingid moth	*Oxyambulyx sericeipennis* Butl.	Jangali badam	*Stercularia foetida* L.
37.	Sphingid moth	*Rhopalopsyche nycteris*	Wild grape	*Vitis vinifera* Linn.
38.	Sphingid moth	*Rhopalopsyche bifaciata* Butt.	Wild mulberry	*Morus nigra*
Family – Limacodidae (Cochlidiidae)				
39.	Limecodid moth	*Latoia lepida* (Cramer)	Mango	*M. indica*
40.	Limecodid moth	*L. repanda* (Walker)	Almond	*Amygdalis communis* L.
41.	Limecodid moth	*Thosea aperiens* Walker	Tamarind	*T. indica*
42.	Limecodid moth	*Nemata lohor* Moore	Jangali Kel	*M. rosacea*
Family – Hyblaeidae				
43.	Teak defoliator	*Hyblaea puera* Cramer	Teak	*Tectona grandis*
Family – Torticidae				
44.	Torticid moth	*Homona coffearia* (Nietner)	Jamun (Jambhul)	*S. cuminii*
45.	Torticid moth	*Strepsicrates rhothia* Meyrick	Jamun (Jambhul)	*S. cuminii*
46.	Torticid moth	*Ulodemis trigrapha* Meyrick	Bitter orange	*C. arantium*
47.	Torticid moth	*Cydia palamedes* Meyrick	Tamarind (Chinch)	*T. indica*
Family – Pyralidae				
48.	Pyralid moth	*Nephopteryx eugraphella* Ragonot	Mango	*M. indica*
49.	Pyralid moth	*Phycita orthoclina* Meyrick	Tamarind (Chinch)	*T. indica*
50.	Pyralid moth	*Orthaga euadrusalis* Walker	Mango	*M. indica*
51.	Pyralid moth	*Orthaga* sp.	Ber	*Z. jujuba*
52.	Pyralid moth	*Diaphania itysalis* Walker	Anjir	*Ficus carica* L.
53.	Pyralid moth	*Eutectona machaeralis* Walker	Teak	*Tectona grandis*
54.	Pyralid moth	*Dichcrocis punctiferalis* Guenee	Neem	*Azadirachta indica* A.Juss.
55.	Pyralid moth	*Synclera univocalis* Walker	Ber	*Z. Jujuba*
Family – Metarbelidae				
56.	Metarbelid moth	*Indarbela dea* (Swinhoe)	Mango	*M. indica*
57.	Metarbelid moth	*I. quadrinotata* (Walker)	White Mulberry	*M. alba*
58.	Metarbelid moth	*I. tetraonis* Moore	Ber	*Z. jujuba*
59.	Metarbelid moth	*I. theivora* (Hampson)	Mango (Amba)	*M. indica*

Contd...

Table 49—Contd...

Sl.No.	Moth Species		Host Plants	
	Common Name	Generic Name	Common Name	Generic Name
Family – Noctuidae				
60.	Noctuid moth	*Eligma narcissus* Cramer	Maharuk	*Ailanthus excelsa*
61.	Noctuid moth	*Calyptra bicolar* (Moore)	Jamun(Jambhul)	*S. cuminii*
62.	Noctuid moth	*Carea chlorostigma* Hampson	Jamun (Parjambhul)	*Olea diotca* Roxb.
63.	Noctuid moth	*Eublemma abrupta* (Walker)	Mango	*M. indica*
64.	Noctuid mtoh	*E. angulifera* Moore	Tamarind (Chinch)	*T. indica*
65.	Noctuid moth	*Helicoverpa armigera* (Hubner)	Anjir	*F. carica*
Family – Noctuidae				
66.	Noctuid moth	*Achaea janata* L.	Ber	*Z. jujuba*
67.	Noctuid moth	*Parallelia algira* (L.)	Ber	*Z. jujuba*
68.	Noctuid moth	*Prodenia littoralis* Fab.	Teak	*T. grandis*
69.	Noctuid moth	*Bombotelia delatrix* Guenee	Jamun (Parjambhul)	*Olea dioica* Roxb.
70.	Noctuid moth	*B. jocosstrix* Guenee	Mango	*M. india*
71.	Noctuid moth	*Calpe emarginata* (Fabricus)	Bitter orange	*C. arantium*
72.	Noctuid moth	*Selapa celtis* Moore	Anjir	*F. carica*
73.	Mango shoot webber	*Orthaga exvinacea*	Mango	*M. indica*
74.	Fruit sucking moth	*Othereis fullonica*	Mango	*M. indica*
75.	Noctuid moth	*Hamodes indica* sp. nov.	Tamarind	*T. indica*
76.	Noctuid moth	*Hamodes shivajiensis* sp. nov.	Fig	*F. glomerata*
77.	Noctuid moth	*Heloithis nocturna* sp. nov.	Common Bean	*P. vulgaris*
Family – Heliozelidae				
78.	Heliozelid moth	*Antispila anna* Flet.	Kala Jamun (Jambhul)	*S. cuminii*
79.	Heliozelid moth	*A. argosioma* Meyrick	Wild jujuba	*Z. olenoplia* L.
Family – Tineidae				
80.	Tineid moth	*Tinea* spp.	Mango	*M. indica*
Family – Gracillariidae				
81.	Gracillariid moth	*Acrocercops loxias* Meyrick	Jamun (Jambhul)	*S. cumini*
82.	Gracillariid moth	*A. phaeospora* Meyrick	Jamun (Parjambhul)	*Oliea dioica* Roxb.
83.	Gracillariid moth	*A. syngramma* Meyrick	Jamun (Parjambhul)	*O. dioica*
84.	Gracillariid moth	*A. zigonoma* Meyrick	Mango	*M. indica*

Contd...

Table 49—Contd...

Sl.No.	Moth Species		Host Plants	
	Common Name	Generic Name	Common Name	Generic Name
85.	Gracillariid moth	*Oecadarchis* sp.	Tamarind (Chinch)	*T. indica*
86.	Gracilariid moth	*Phyllonorycter epichares* Meyrick	Ber	*Z. jujuba*
Family – Lymantridae				
87.	Gypsy moth	*Lymantria concolar* Walker	Mango (Amba)	*M. indica*
88.	Lymantrid moth (Hairy catterpiller)	*Euproctis flava* (Bremer)	Mango (Amba)	*M. indica*
89.	Lymantrid moth	*E. lunata* Walker	Ber	*Z. jujuba*
90.	Lymantrid moth	*Lymantria ampla* Walker	Mango (Amba)	*M. indica*
91.	Lymantrid moth	*Porthesia scintillans* (Walker)	Mango (Amba)	*M. indica*
Family – Lasiocampidae				
92.	Lasiocampid moth	*Streblote siva* (Lefebvre)	Ber	*Z. jujuba*
93.	Lasiocampid moth	*Trabala vishnou* (Lefebrre)	Parjambul Jamun	*Olea dioica* Roxb. *S. cumini*
94.	Lasiocampid moth	*Malacosoma indica* (Walker)	Hirda	*Terminalia chebul*/Retz.
95.	Lasiocampid moth	*Suana concolar* (Walker)	Bitter orange	*C. arantium*
Family – Syntomidae				
96.	Syntomid moth	*S. hampson*	Karanj	*Pongomia pinnata*
Family – Psychidae				
97.	Psychid moth	*Acanthopsyche minima* Hampson	Banana (Jangali kel)	*Mangifera rosacea*
98.	Psychid moth	*Acanthopsyche* sp.	Mango (Amba)	*M. indica*
99.	Psychid moth	*A. plagiophleps* Hampson	Tamarind (Chinch)	*T. indica*
100.	Psychid moth	*Chalides vitrera*	Tamarind (Chinch)	*T. indica*
Family – Saturnidae				
101.	Moon moth.	*Actias selene* (Hubner)	Wild Cherry	*Prunus avium* L.
102.	Tassar moth	*Antheraea mylitta*	Ain Ber	*T. tomentosa* *Z. jujuba*

In all, 103 species of moths have been reported from western Ghats of Kolhpur and Satara district of Maharashtra. Out of which 84 were common and abundant. 6 species were rare and 13 species were found new to science in the region. The rare species of moth were *Hippotion volax* from family Sphingidae, *Orthaga euadrusalis* Walker, *Daphania itysalis* from family Pyralidae. *Bombotelia delatrix* Guenee from family Noctuidae, *Antispila anna* from family Helizelidae and *Actias selene* (Hubner) from family Saturnidae.

7

Concepts of Pest Lepidoptera

The new species described in the book refer to *Argina kolhapurensis* sp.nov., *Argina indica* sp.nov., *Argina satarensis* sp.nov., *Spilosoma sathei* sp.nov., *Pelochyta sathei* sp.nov., *Moorea indica* sp.nov., *Moorea marathi* sp.nov., *Cephanodes indica* sp.nov., *Dilephila indica* sp.nov., *Hamodes shivajiensis* sp.nov., *Hamodes indica* sp.nov., *Heliothis nocturni* sp.nov. and *Hypsa marathi* sp.nov. The above species have been compared with the closely related species described by Hampson (1976), and new and earlier described species have been confirmed. In the text ten species have been redescribed which refer to *Argina guttata* (Hampson), *Creatonotus gangis* (Linnaeus), *Rhoplopsyche nycteris*, *Rhoplopsyche bifaciata*, *Theretra nessus*, *Eligma narcissus* (Crammer), *Helicoverpa armigera* (Hubn.), *Prodenia littoralis*, *Hypsa producta* Moore and *Eutectona machaeralis* Walker. While redescribing the species, additional characters related to head, thorax and abdomen have been added. The identifying keys for subfamilies, genera and species have been incorporated. The taxonomical work will be helpful for easy identification of species and protection of forests and agricultural crops from western Maharashtra including Ghats.

Beeson (1941) reported that oviposition period in *E. macheralis* ranged from 7 to 14 days. Patil and Thontadarya (1987) reported preoviposition and oviposition periods as 4 to 7 days and 4 to 11 days respectively. Similarly, David and Kumarswami (1982) reported 250 to 500 eggs laid by the females while Patil and Thontadarya (1987) reported 205 to 348 egg laid by the female *E. machacralis*. According to Beeson (1941) eggs averaged 0.75 mm in length.

Stabbing (1903) studied incubation period in *E. machaeralis*. He reported 3 to 4 days as incubation period for the eggs. Maximum eggs were hatched on third day while, David and Kumarswami (1982) recorded 2 to 3 days incubation period in *E. machaeralis* but, Patil and Thontadarya (1987) recorded 3 to 4 days as incubation period in the laboratory.

The larval development was completed from 12 to 20 days during which *E. machaeralis* passed through five instars (Beeson, 1941). David and Kumarswami (1982) reported 18 to 27 days as larval period. While, Patil and Thontadarya (1987) reported 12.0 to 19.3 days as larval developmental period in the same species. The present findings were in agreement with those of Beeson (1941), who reported that the larval period was 12 to 20 days at Nilambur, Chennai and 12 to 19 days at Pyrimana, Myanmar. David and Kumarswami (1982) reported 18 to 27 days as larval developemental period while, Patil and Thontadarya (1987) noticed five larval instars and the larval periods for first, second, third, fourth and fifth instars were 2.5 to 4.0, 3.0 to 4.0, 2.5 to 4.0, 3.0 to 4.0 and 4.0 to 5.0 days respectively.

According to Beeson (1941) pupal stage lasted for 6 to 7 days in *E. machaeralis* while, Patil and Thontadarya (1987) reported pupal period as 7 to 8 days during July – August, 8 to 9 days during December – January and 6 to 7 days during March – April, in the same species.Beeson (1941) studied wing expansion of male and female moths in *E. machaeralis* as 21 to 26 mm and 19 to 24 mm respectively and adult longivity as 9 days, while Patil and Thontadarya (1987) reported longivity of *E. machaeralis* male and female moths as 8.80 to 15.45 days and 11.90 to 19.95 days respectively. In present study female moths of *E. machaeralis* survived for 6.9 days when fed with 5 per cent sugar solution. According to Beeson (1941) the incubation, larval, prepupal, pupal and adult stage varied from 2 to 3 days, 8 to 27 days, 1 to 2 days, 4 to 11 days and 9 days respectively in *E. machaeralis*. He also reported that the life cycle was completed in 24 to 41 days, similarly, Fletcher (1914) reported 30 to 32 days as total life cycle period while, Patil and Thontadarya (1987) recorded 24.2 to 38.1 days for completion of total life cycle from egg to adult in *E. machaeralis*. Beeson (1941) counted sex ratio (male : female) as 1:1.94 in *E. machaeralis* females outnumbered the males. He reported that female moths were often twice as much as male moths in number while, Patil and Thotadarya (1987) reported that the percentage of male to female ratio was favouring males which is contrary to the present findings.

Joseph and Karnavar (1991) studied identification of sex in the larval and pupal stages of *Ailanthus* defoliator, *E. narcissus*, the major monophagous pest of *Ailanthus* in Kerala. The larvae cause serious damage by feeding and defoliating young and mature leaves. While studing the morphogenetic process of this insect they felt the need for an accurate and quick method for sexing the pupae. Pupal sexing has been reported earlier in a number of lepidopterous insects based on various morphological and morphometeric characters (Mosher, 1916; Butt and Cantu, 1962; Solomon, 1962; Narayanan *et al.,* 1977; Navaranjan paul *et al.,* 1979; Qureshi *et al.,* 1986; Santosh Badu and Prabhu, 1887; Annie John and Muraleedharan, 1989).

More frequent abdominal wriggling by male pupae in *Ennomous subsignarius* has been reported (Solomen, 1962). Abdominal wriggling was observed in both sexes but, male showed more frequent with active abdominal movements. Male pupae were comparatively smaller in size than female and gave fair degree of accuracy in sexing laboratory reared insects.

Mass rearing technique of *S. litura* (Fabricius), a polyphagous noctuid of much economic importance has been developed by Shorey and Hale (1965), Chu *etal.,*(1976),

Boardmen (1977), Okada (1977) and Gupta and Pawar (1985). All of them have reared the larvae individually on semi-synthetic diet in glass vials.

Raj (1988) studied seasonal variation in the male population of potato tubermoth, *Phthorimaea operculella* (Zell.) in deccan plateau. *P. operculella* is one of the most destructive pests of potato causing severe damage to the plants as well as the tubers. The larva mines the foliage, stem and tuber. In India, 30 to 70 per cent tubers are infested under different indigenous methods of storage (Lal, 1949, Sen, 1954 and Wesley, 1956). Although the life cycle of this insect pest has been studied in many countries (Mukharjee, 1949, Nirula, 1960, Staner and Kaitazov, 1962, Doreste and Nieres, 1968, Broodruk, 1971, Al-Ali *et al.,* 1975, Foot, 1979 and Gomma *et al.,* 1979) no detailed study have been made in relation to adult sexual dimophism and eye pigmentation. Raj (1988) studied following aspects of life history of *P. operculella* such as hatchability of eggs, number and duration of different larval instars, duration of different pupal stages based on the eye pigmentation, sexing of adults, fecundity rate of female adults and longevity and sex ratio of adults. The exact number of larval instars was calculated by measuring the width of head capsule at different developmental stages. He reported among different life stages of potato tuber moth, larva was the only destructive stage, which caused damage to potato tubers. Damage was minor during 1 to 3 days after infestation but, became severe in the next 8 to 9 days. The longivity of male adult was more than the female moth. The ratio of male to female was 1:1.1. Gubbaiah and Thontadarya (1977) studied the binomics of *Gnorimoschema operculella* Zeller under field conditions. Female moth laid an average of 208 ± 27.4 eggs during life span and the maximum number of eggs were laid on the 3rd day after emergence. For the act of oviposition, female moth prefered rough surface, as maximum number of eggs were laid on muslin cloth than on smooth surface.

Mathew *et al.* (1990) studied an artificial diet for the teak defoliator *H. puera*. An artificial diet containing teak leaf powder, Kabuligram flour and other commonly available ingredients was developed for rearing the teak defoliator *H. puera*. The basic composition of the diet used in experiment was similar to that developed by Nagarkatti and Prakash (1974) for *H. armigera*. The diet may be taken as the base diet and improvement attempted. Genaral observations showed that storing the diet for a period of 2 to 3 months in a refrigerator had no deleterious effect on survival of the insects.

Biology, infestation characteristics and impact of the bagworm, *Pteroma plagiophleps* Hampson; in forest plantations of *Paraserianthes falcataria* have been studied by Nair and Mathew (1992). Their results indicated that the bagworm *P. plagiophleps* until recently an important insect associated with the tamarind tree, has emerged over the last 10 years as a serious pest of new forest planting of *P. falcataria*. One of the major characteristics of *P. plagiophleps* infestation was that population outbreaks occured only in some plant species although it has a wide host range covering diverse families. Site factors play a decisive role in the population dynamics of many insects (Berryman, 1986). Forest insect outworks can be prevented by site amelioration (Beryman and Baltensweiler, 1981). Few studies have been attempted on population dynamics of bagworms, a group of insects with peculiar biological attributes and containing many important pests of tree crops in the world. (Mathew and Nair, 1984).

Dumbre *et al.,*(1989) studied Tasar Silkworm rearing in the Konkan area. They reported that availability of tasar food plants were abundant and is one of the prerequisites for tasar culture. Ain trees (*Terminalia tomentosa*) was a good food plant for tasar silk worms. Ain trees are plentiful in the Konkan region. The authors hopefully conducted the experiments to explore the possiblilities of introducing tasar silkworm rearing on a commercial scale in the Konkan region using Ain trees as food plants. In the present study, *E.machaeralis, E. narcissus* and *H. producta* were selected for biological studies since they are destrictive insects of forest flora.

Genital morphology of 16 species of Pyraustinae belonging to 14 genara were studied and modification of the various parts in the different species were summerised. The pyraustinae form an economically important group of Lepidoptera characterised by the presence of a well developed proboscis and the free origin of the forewing and vein 7 and 10 from the cell. Recent studies by several workers (Munroe 1960, 1969, 1970; Pajni and Rose, 1977; Mathew and Menon, 1986 a, b, c) have shown the importance of genitalic structures in the classification of this group of insects.

Suryawanshi *et al.* (2001) studied life tables for *Earias vitella* (F.) on different hosts in the laboratory. They indicated that depending on survival values, the most suitable food for this insect was okra seeds (0. 89), followed by okra fruits (0.80), cotton squares (0.73) and cotton bolls (0.68). The net reproductive rate (R0) was highest in okra seeds. (141.50) followed by okra fruits (124.17), cotton boll (100.62) and cotton squares (99.32). The mean length of generation (T) ranged from 32.51 on okra seeds to 38.48 days on cotton square. The highest value of innate capacity for increase in numbers (rm) was 0.1523 on okra seeds and lowest, 0.1195 on cotton squares, on the bases of r_m the food plants suitable for the pest were okra seeds (9.1523), okra fruits (0.1385) cotton bolls (0.1523), okra fruits (0.1385), cotton bolls (9.1289) and cotton square (0.1195), in descending order. On reaching stable age distribution the population of *E. vitella* at egg, larva, pupa and adult stages comprised 53.33, 39.59, 6.07 and 0.92 per cent on okra fruits, 54.99, 37.08, 6.96 and 0.93 per cent on okra seeds, 51.72 41.55, 5.69 and 0.98 per cent on cotton boll and 49.05 43.76, 5.60 and 0.92 per cent on cotton squares respectively. In the present study attempts have been made on biology, sex ratio and fecundity. The data is essential for construction of life table statistics. Life table studies have been carried out in the present forms *i.e. E. machaeralis E. narcissus* and *H. producta,* will be helpful for their control strategies.

Kalia and Harsh (2003) studied a fungus *Metarhizium anisopliae* Anthn against the larvae of teak defoliator, *H. purea.* It showed pathogenicity to all the larval instars of teak defoliator. According to them several species of fungi were known to cause death in insects *M. anisopliae* was used for controlling pasture inhabiting insects (Latch, 1965) (Veen, 1968). In western Ghats fungi species are more abundant than plain region. Entomogenus fungi will be investigated in future. Bhalani (1989) studied the suitability of host plants for growth and development of leaf eating caterpillar *S. litura.*

The leaf eating caterpillar, *S. litura* is polyphagous pest and widely distributed throughout the India. Its biology and feeding response and suitability to different hosts were studied by some research workers (Basu, 1944; Thobbi, 1961; Ratan Lal and Nayak, 1965; Bhilani and Talati, 1984 and Patel *et al.,* 1987). However, the degree of host plant suitability was functional factor for the insect including its nutritional composition.

Thorsteinson (1960), Yamamoto and Fraenkel (1960) studied the suitability of host crop plants for growth and development of *S. litura*. The results on the differences in the larval period of *S.litura* on different hosts were significant. The leaf eating caterpillar completed its larval period in a shorter time in the highly suitable host (castor) rather than unfavourable host (maize). The next shorter period was observed on cowpea host, whereas, the larval period was almost equal on the rest of the host plants. However, a marked variation in the duration of larval development was occurred when the rearing was done on different food plants (Basu, 1944; Ratan Lal and Nayak, 1963). It was evident that those larvae which maintained better growth by feeding on a particular food plants had comparatively shorter larval period. Similar findings have been reported by the workers (Ratan Lal and Nayak, 1963; Bhalani and Talati, 1984).

Patel *el al.* (1987) reported growth index for castor 5.75. Their results were in conformity with those attained by other workers (Teodia and Singh, 1968; Bhalani and Talati, 1984). Data on the pupal length of *S. lilura* revealed that the differences in size were apparent among the pupae reared on the seven food plants and variations between them were highly significant. In general, larvae reared on favourable host were greater in pupal size. It would be very interesting to test above aspect in present forms.

Painter (1968) who was of the view that, the food plant of the immature forms determines the fecundity of the adult insects. Slow growth and development, high mortality of immature stages and low reproduction in adults were the familiar symptoms of nutritional deficiencies (House, 1963). These finding suggest that the nutritional factors play an important role in the better suitability of the host for the growth and development of *S. litura*. In the present study, fecundity was 151.8, 157.9 and 187.4 in *E.narcissus, H. producta* and *E. machaeralis* respectively.

Masoodi and Srivastava (1985) studied effect of host plants on the pupal weight and fecundity of *Lymantria obfuscata*(Walker). *L. obfuscata* is one of the serious pest of fruit and forest plantation in Kashmir (Malik *et al.,* 1972). The sudden increase in its area of infestation and host range has posed a serious threat to the fruit industry (Shaikh, 1975). Similarly, in *Lymantria dispar* L. some correlations between the diet and pupal weight and fecundity have been established (Zecevic, 1958; Kansu, 1962; Merker, 1964; Hough and Pimental, 1978; Vasov, 1979; Barbosa and Green blat, 1979).

Javed Iqbal Siddigi (1985) studied *Diacrisia (= Spilosoma) obliqua* Walker with respect to damage to crops and distribution. It is an important polyphagous pest in India. The larvae are commenly called "Bihar hairy caterpillars. The caterpillars were omniphagous and voracious feeders and have been reported practically from all over the country to cause considerable damage to pulses, cotton, vegetables, sorghum, maize, rags, oil seeds, small millets, sugarcane, paddy, wheat, guinea grass, jute, sun hemp, beet-root, potato and sweet potato. Besides, they also attack other agricultural and fiber crops as well as medicinal plants (Pant, 1964). The pheromonal communication was recorded earlier in this species (Siddigi, 1982, 1985). Pheromonal studies in the present investigations were not carried out. However, these aspects will help in adopting ecofriendly control measures in future.

Masoodi (1985) studied growth response of *L. obfuscata* in relation to tannin content in different host foliages. Food was one of the important determinants of population dynamics of phytophagous insects (Painter, 1936) and the biochemical profile of the host plant affects growth, survival and reproduction of insect in various ways (Schoonhoven, 1968). Besides, lower water and nitrogen content, some secondary compounds provide a degree of protection for plants from insect damage.

Madhuchandra and Chaudhuri (1992) studied the archid moth *Diacrisia casignetum* Kollar, a pest of sunflower (Banerjee and Haque, 1983) and castor (Chatterjee and Chaudhari, 1990) have also been found to attack other plants as well. Sixteen plants obtained from the natural habitat of *D. casignetum* were tested for their acceptance to this insect, of those ten plants accepted within the first three hours by the caterpillars for growth and development. The host plant showed differential reaction to food consumption weight of larvae, rate of larval development and adult emergence. Castor was found to be the most suitable plant for larval development and recorded the highest growth index value while hempweed the least. There was an extra moult on mulberry. The highest percentage of adult emergence was found in sesame while, least in hempweed.

Tiwari *et al.*(1988) reported susceptibility of several varieties of groundnut to Bihar hairy caterpillar *S. obliqua*. Occurance of this pest on groundnut *Arachis hypogaea* L. was reported by several workers (Lefroy, 1907; Andres 1915; Fletcher, 1917; Srivastav *et.al.* 1963). Pandey *et al.* (1968) indicated that this insect can successfully develop on groundnut. In recent years, a large variety of prereleased varieties and germplasm of groundnut are under trial for improvement of agronomic characters.

Goyal and Rathore (1988) studied patterns of insect plant relationships on the basis of susceptibility of different hosts to *H. armigera* called gram pod borer, a cosmopolitan, polyphagous pest which seriously damage gram crop in Madhya Pradesh and different parts of India. The susceptibility to different host plants to a pest depends on the insect plant relationships. Hence, they worked on orientation, fecundity, survival, growth and development by the food intake, nutritive value of food, and physical and chemical characteristics of the plant (Saxena, 1969). Such ecological studies may give clues for evolving pest management systems. Ecological studies for the present forms are needed.

Singh and Dhamdhere (1989) studied field screening of some guava varieties against the bark eating caterpillar *Indarbela quadrinotata* (Walker). Gauva *Psidium guajava* (L.) is one of the many host plants of bark eating caterpillar *I. quadrinotata*. Occasionally, it occured in large number causing severe damage especially in neglected and unclean orchards. Older trees were more prone to their attack than the young ones (Butani, 1979). The caterpillars consumed the bark and produced a web like structure interspread with its excreta, sawdust etc. Later, the caterpillar bored into the trunk for feeding. Excessive feeding of the caterpillars caused interruption of cell sap which adversely affected the growth and fruit setting capacity of the tree. Normally, one larva was seen inside a bore hole. Little work appears to have been done on varietal screening of guava against this pest excepting that of Sandu *et al.* (1977) and Batani (1979). None of the varieties evaluated was found to be free from the attack of the bark eating caterpillar. Taking an overall picture it was inferred that seedless, apple gauva and Allahabad safeda are more susceptible to the attack of the pest. This is in conformity with the finding of Butani (1979) in respect of seedless guava and apple but differ in respect of Allahabad safeda which has been reported to be resistant to the attack of this pest. They reported that the older trees were found more infested than young ones. Similar observations were made in the past by Sandhu *et al.* (1977), Butani (1979) and Singh (1984).

Singh and Sehgal (1993) studied the consumption and utilization of different food plants by larvae of *S. obliqua*. Bihar hairy caterpillar *S. obliqua* is a versatile, widely distributed polyphagous insect pest causing damage to a broad range of plants of agriculture and medicinal importance. It is serious sporadic pest in oriental region. A polyphogous insect does not damage all the host plants with equal severity. There is considerable speculation for such selection of food by insect. Information on consumption and utilization of food by insect is important in physiology, nutrition, ecology and economic entomology (Waldbauer, 1964). Each insect species is adapted to a specialized food which they can utilize more efficiently. The qualitative nutritional differences have to be sought at quantitative level. Hence, a clear picture of comparative nutrition of insects does not emerge until quantitative studies are emphasized (Premkumar *et al.,* 1977). For better understanding of the nature of food material, it is obligatory to collect information on the rate of feeding and its effect on growth and development, the amount of food digested and the quantity of food converted into body mass (Sanjayan and Murugan, 1987).

Tiwari (1985) studied effect of Dimilin on consumption and utilization of dry matter and dietary constituents of castor *Ricinus communis* (Linn.) by the castor semilooper *A. janata* Linn. The effect of various insecticides on the feeding and utilization of dry matter and dietary constituents by insect has been reported from time to time (Ramdeo and Rao, 1979a, b, c; 1980; Turunon, 1977; Vishwa Nath, 1983; Nath and Rao, 1983). Dimilin (diffubenzuron) a derivative of urea is known to be an inhibitor of chitin synthesis in insects. On the larvae of *A. janata* Dimilin as a juveno-mimetic compound has already been established by Subrahmanyam *et al.* (1980). Except above workers so far no study has been undertaken on dimilin effect on consumption and utilization by insects.

Prem Sagar *et al.* (1990) studied field screening of different jute cultivars against Bihar hairy caterpillar *S. obliqua* in Punjab. Jute is an important fibre crop in India. A large area was brought under the cultivation of jute and other blast crops after partition to meet the requirements of a large number of mills which would have remained idle otherwise and consequently the country was to suffer a heavy loss of foreign exchange.

It has two distinct cultivated species *viz., Corchorus olitorus* L. and *C. capsularis* L. which are grown mainly in the deltaic and river basins of the eastern states of the country namely, Assam, Bihar, Orissa, Tripura, Uttar Pradesh and West Bengal. In Punjab, jute was not popular crop but, efforts were made to introduce it. The crop suffers from the degradation of large number of insect pests. Lefroy (1906; 1907) reported Bihar hairy caterpillar *S. obliqua* as minor pest. Kundu (1956) briefly mentioned about a few pests of jute.

Sharma and Tara (1988) studied comparison of consumption and utilization of mulberry leaves in two moths *S. litura* and *D. (= S.) obliqua*. Both are known to be polyphagous pests (Aitkenhead *et al.,* 1974; Patel *et al.,* 1973; Singh and Byas 1973; Basi, 1972). Work on the rate of intake of food, growth, digestibility and rate of consumption and utilization of food by these insects has been done by Bhatt and Bhattacharya (1978) and Babu *et al.* (1979) but, these investigators have used soybean and green gram as the host. Tara (1983) studied these parameter for both the pests in relation to mulberry leaves as these have been observed damage to plantations in Jammu region. Sharma and Tara (1985) realized the importance of such studies in determining insect host relationship and the extent of damage by above pests to mulberry plantations.

Singh and Bhattacharya (1994) studied formulation of semisynthetic diets for Bihar hairy caterpillar, *S. obliqua*. Bihar hairy caterpillar is a polyphagous pest, attack extensively a large varieties of cultivated as well as non–cultivated plant species (Deshmukh *et al.,* 1976). The development of suitable artificial diet of insect provides a good understanding about the relationship between insect and plants, such knowledge would help in biological, nutritional biochemical, host plant resistance and toxicological studies of insects which will be ultimately used in developement of safer and economical pest management programms. In past, fundamental and applied fields of entomology were severely affected due to lack of suitable artificial diets for insects (Knipling, 1966; Pant, 1973; Vandorzans, 1974 and Singh, 1977). Earlier, Tiwari and Bhattacharya (1987) formulated several semisynthetic diets for insects. However, they used only ten different stored commodities, may greatly alter the quality of the diet. Therefore, it is advisable to formulate the semisynthetic diets for insects by using some other commonly available commodities which were not used earlier.

Singh and Singh (1991) studied emergence pattern of pink boll worm moths from overwintering larvae. In north India, large portion of larvae of pink bollworm *Pectinophora gossypiella* (Saund.) enter into diapause due to fall in temperature and low photoperiod. The diapausing population act as an off season carryover of pest from one year to another. A portion of the pink bollworm overwintering population emerge before cotton begins to fruit. Their emergence is commonly referred to as suicidal because no food and habitat for the newly hatched larvae are available on the cotton plants. But, the adult emerged after the month of May are of economic importance which cause economic damage to the cotton crop latter in the season. The last date of suicidal emergence depends upon temperature and the growth stage of the host plants (Bariola, 1978). The adequate knowledge about the emergence pattern of moth from hibernating larvae may be helpful to evolve the suitable techniques for the control of *P. gossypiella,* a notorious pest.

Khan and Srivastav (1993) studied biological interacton of Penfluron against *Euproctis icilia* Stoll. Penfluron belongs to a new class of chemical generally known as insect growth inhibitor which causes lethality in immature stages by inhibiting the chitin biosynthesis during moulting, produces abnormality in developemental stages due to defects in process of cuticle deposition and induces sterility in surviving adults due to disruption of cuticle formation in the developing embryo (Grosscurt, 1978). Penfluron has been used by several workers for the pest management. Borkovec *et al.*,(1978) evaluated a procedure suitable for sterilizing both sexes of *Anthonomous grandis*. They also observed absolute mortality at 0.05 g/litre level of treatment, Study was carried out in the laboratory to find out the suitability or penfluron against the population management of *E. icilia* which is serious pest of castor crop (*R. communis*).

Samarjit *et al.*,(1992) studied distribution pattern of diamond black moth (*Plutella xylostella* L.) on cabbage and cauliflower. Diamond black moth *P. xylostella* is a serious pest of crucifer vegetables and becoming a limiting factor in growing cabbage (*Brassica oleracea* L.) *Convar capitala* (L.) *Alef. var capital* (L.) and cauliflower (*B. olevacea* (L.), *Convar botrytis* (L.), *Alef var botrytis* (L.) crops. Extensive work has been done on different aspects of the pest (Chelliah and Srinivasan, 1987; Feng, 1987; Lin *et al.*, 1988). Spatial distribution is one of the most characteristic ecological properties of a species (Taylor, 1984). It provides reliable estimate of field population densities which is primary need in a pest management programme. They determined the spatial distribution of the larval stages of this pest in farmer's field under different spray conditions.

Roychoudhary *et al.* (1991) studied growth, fecundity and hatchability of eggs of *B. mori* L. in relation to rearing space. An experiment was carried out in the plains of West Bengal following Japanese, Indian and Chinese spacing schedule for bivoltine silkworm *B. mori* to find out growth of different developmental stages, number of eggs laid and hatching performance of bivoltine hybrid 'NB'$_{16}$ × 'P'$_5$. Result revealed that among the spacing as per Chinese recommendation played a significant (P < 0.05) and positive role for improvement of larval, pupal and imaginal weight, fecundity and hatching performance of eggs, whereas crowded condition was found to be deterimental reflected by Japnese spacing schedule. Indian spacing schedule exhibited the performance in between Japanese and chinese except for the hatching potentiality.

Mathew *et al.*,(1990) studied an artificial diet for the Teak defoliator *H. puera*. An artificial diet containing teak leaf powder, Kabuligram flour and other commonly available ingredients was developed for rearing the teak defoliator *H. purea*. Growth and development of 3-4 days old larvae (initially established on teak leaf) on two diet combinations was compared with that on teak leaf. One of the diets was found to be better than teak leaf with respect to percentage, survival and pupal weight.

Intrinsic Rate of Increase

Bains and Shukla (1976) studied the life tables and intrinsic rate of increase in *Chilo partellus* (Swin.) (lepidoptera), the intrinsic rate of increase (r_m) at different temperatures were in ascending order 0.0002 (35°C), 0.165 (32.5 °C), 0.223 (25°C), 0.383 (27.5°C) and 0.435 (30°C). These conclusions showed that the rate of increase was maximum at 30°C which should be considered to be the optimum temperature for the multiplication of this lepidopterous pest. However, the present study was not carried out at different temperature. Further observations of Bains and Shukla (1976) on the finite rate of increase per week were 4.67, 15.59, 21, 3.177 and 1.002 at 25°C, 27.5°C, 30°C, 32.5°C and 35°C respectively, In the present study l was calculated for each lepidopterous pests (*E. narcissus, E. machaeralis, H. producta*) in respect of daily increase at laboratory temperature (25 ± 2°C, 65 ± 5 per cent R.H. and 12 hr photoperiod.).

In *H. armigera,* the value of R_0 indicated that 285.06 females were produced per female during one generation. The innate capacity and finite rate for increase in numbers were 0.1210 and 1.1260 respectively. The mean duration of a generaton was 46.71 days. Under conditions of abundant space, the daily finite rate of increase of *H. armigera* was 1.1286 which enabled the insect to multiply 2.3322 times every week (Bilapate and Pawar, 1980).

According to Reddy and Bhattacharya (1988) the life expectancy (e_x) of *H. armigera* declined up to first 6 days due to egg mortality and increased upto 10th day due to larval mortality. Later, with the advancement of development e_x decreased steadily till it reached 46th day. This type of enhancement in e_x due to heavy mortality at any age group was also reported for *Naranga diffua* Walker, *Phyllonistis citrella* Stainton, *Creatonotus gangis* Linneaus, *S. obliqua* and *S. litura* (Singh, 1984). There was indication of the survival fraction (lx) of each cohort. Females started laying eggs after 31.5^{th} day and stopped it after 39.5^{th} day with lx values being 0.42 and 0.17 respectively. The lx decreased gradually after 4.5^{th} day due to adult mortilty.

Fecundity rate (mx) and reproductive rate (*lx* mx) of each age group showed an undulating pattern during the entire egg laying period. Such pattern was also reported for several other insects (Evans and Smith, 1952; Choudhary and Bhattacharya, 1986). Reddy and Bhattacharya (1988) studied various life parameters computed to get an overall picture of different vital statistics of *H. armigera* on maize based diet. Mean length of generation (T) indicated that this insect completed first generation in 35.5 days. Similarly, net reproductive rate (R_0), accurate estimate of intrinsic rate (r_m), finite rate of increase or the population multiplication in on unit time (l), time required for the population become double (DT), potential fecundity (PT) and annual ratio of increase (AR) were 46.98, 0.1090, 1.1152, 3.36, 134.40, 1.898×10^{17} respectively.

In the present study 'r_m' and 'T' of *E. narcissus, E. machaeralis* and *H. producta* were 0.129, 33.54; 0.149, 30.06 and 0.141 and 30.78 days respectively. The present studies will be helpful for population dynamics of above forest pests and in deciding control strategies for them.

8

Pest Lepidoptera Control

Lepidopterous insects are characterized by having scales on their body. Butterflies and moths are grouped under the Order - Lepidoptera. This is very large order containing more than 1, 50,000 species. Many butterflies and moths are brilliantly colored. Lepidopterans may be small sized to very large sized. They are provided with coiled suctorial proboscis. They may feed on nectars and fruits juice in adult stage. There are four distinct stages of their life namely, egg, larva, pupa and adult. The larvae are mostly phytophagus but, some also feed on cloth, wax, honey, Cadburys, etc. and some ones feed on other insects and acts as predators. However, some useful lepidopterans are well known like silkworms on which the sericulture industry is based.

Lepidopteran insects mostly cause the damage to agricultural crops or other products in larval stage. Pupa doesn't feed and not destructive to any product. However, some adult lepidopterans can cause the damage to products like citrus fruits and honey combs. The larvae of Lepidoptera are very destructive to agricultural, floricultural, horticultural and forest crop plants. They feed on leaves and act as defoliators or leaf miners. They also cause the damage to stem, roots, fruits and seeds by boring and difficult to control with conventional pesticides. Because, they enjoy internal or sedentary life in the tunnels. The lepidoterous larva is called as caterpillar. It is characterized by having 3 pairs of walking (true) legs and few pairs (1 to 6 pairs) of prolegs or pseudolegs. The caterpillar shows well developed head, 3 thoracic and 10 abdominal segments. The prologs/pseudolegs are situated on ventral side of the caterpillar and present on 3-6 and 10th segments. In semiloopers pseudolegs are reduced to 1-3 pairs hence, they form loop when they walk.

Lepidopterans are very famous by pest insects in the world including India. In tropical countries lepidopterans are several hundreds as pest insects. In India, hundreds of lepidopterans are major insect pests of agriculture and forest plants. Lepidopterans are worst enemies of forest trees which cause serious damage to the leaves, stems, roots, fruits and seeds of the forest trees. Noctuids, pyraustids, pyralids, lymantrids, and many microlepidopteras are largely associated with forest trees which cause serious treat to forestry. Therefore, their control is essential part of conservation and protection of forest plant biodiversity. Secondly, forestry and biodiversity plays an important role in sustainable development of a country or a region.

More important lepidopterous pests of forest trees are mentioned in Table 50.

The control aspects of lepidopterous pests are divided into two categories.

1. Preventive control measures, and
2. Curative control measures.

Preventive Control Measures

Preventive control measures of lepidopterous pests are given below.

1. Collection and destruction of life stages of the pests such as eggs, larvae, pupae and adults. *L. mathura* lay eggs in masses, it is possible to collect and destroy its egg masses.
2. Collection and destruction of infested parts of crops such as leaves, seeds, fruits, bark, flowers etc. along with pest stages. *Natada velutina, L. mathura, P. reflexa* etc. can be controlled by this way in forest ecosystem.
3. Hand picking of caterpillars and destruction of them. This is possible in pests which have gregarious instars, specially first and second instar. However, other instars also possible to collect by hand picking by keeping labour.
4. Use of sticky bands around tree trunk can help preventing asending of caterpillars on the trees. This is specially possible in *L. mathura*.
5. Many defoliating pest larvae descend down on the ground for pupation. Therefore, ploughing and digging the soil help to expose pupae to natural mortality factors such as high intensity of temperature and light, predators and parasitoids, etc. Some species of pests pupate on plants or among fallen leaves. The pupae should be collected and destroyed and clean cultivation plays important role in such cases. *T. vishnou, P. reflexa* etc. can be controlled by this method.
6. Adults are attracted to light. Hence, collected by light traps and destroyed.
7. For bark feeders scraping the bark of the tree with wooden knife will remove caterpillars residing in cracks and crevices of bark.

In *H. puera* pupation takes place in a triangular leaf fold or in entangled leaf parts. They should be collected and destroyed. Similarly, pupation takes place on green leaves or on fallen leaves of *T. grandis* in a thick shelter web spun in two layers by *H. machaeralis* which are criss-crossed. In such cases pupae should be collected along with plant debris and destroyed or burnt.

Table 50: Lepidopterous Pests of Forest Trees and their Damage

Sl.No.	Name	Host Plant	Damage
1.	Cosmopteryx bambusae (Cosmopterygidae)	Bambusa sp., Dendrocalamus	Larvae mine leaves
2.	Indarbela tetraonis (Walk.) (Metarbelidae)	Acacia sp., Khair	Bark feeder and bore stem
3.	Taragama siva (Lef.) (Lasiocampidae)	Acasia sp.	Acts as defoliator
4.	Metanastria hyrtaca (Cr.) (Lasiocampidae)	Babul, Khair	Acts as defoliator
5.	Euproctis lunata (Wlk.) (Lymantriidae)	Babul, Khair	Acts as defoliator
6.	Clania crameri (West.) (Psychidae)	Babul, Khair Casurina	Acts as defoliator
7.	Arthroschista (Margaronia) hilaralis (Pyralidae)	Kadamba	Larva feed on leaves, defoliator
8.	Atteva fabriciella (Yponomeutidae)	Maharukh Ailanthus excelsa	Larva feed on leaves, defoliator
9.	Hypsipyla robusta (Pyralidae)	Mahogany Swietenia macrophylla Toon	Larvae bore into shoot, Larvae bore into shoot
10.	Ingura subapicalis (Noctuidae)	Sal Shorea robusta Teak Tectona grandis	Larvae feed on leaves Larvae defoliators
11.	Lymantria mathura (Lymantriidae)	Sal Shorea robusta	Larvae defoliators
12.	Trabala vishnou (Lasiocampidae)	Sal Shorea robusta	Larvae defoliators
13.	Pammene theristris (Eucosmidae)	Sal Shorea robusta	Larvae feed on ripening seeds, bore shoots, seedlings
14.	Eupterote undata (Eupterotidae)	Semul Bombax ciba	Larvae defoliators
15.	Tonica niviferana (Oecophoridae)	Semul Bombax ciba	Larvae bore into shoots.
16.	Natada velutina (Limacodidae)	Semul Bombax ciba	Larvae defoliators
17.	Hyblaea puera (Hyblaeidae)	Teak T. grandis	Larvae defoliators
18.	Hapalia machaeralis (Pyralidae)	Teak T. grandis	Larvae feed on leaf epidermis
19.	Phassus malabaricus (Hepialidae)	Teak T. grandis	Larvae bore into saplings
20.	Dichcrocis punctiferalis (Pyralidae)	Teak T. grandis	Larvae damage fruits
21.	Creatonotus transiens (Arctiidae)	Toon Cedrela toona	Larvae defoliators
22.	Agrotis ypsilon (Noctuidae)	Forest nursery	Larvae cut seedlings

In borers pupation takes place in tunnels, at the base of the tunnel. This is noticed in *H. robusta* and *P. malabaricus.* In case of *Tonica niviferana* pupation takes place either on the leaves or stems. Those pupae be collected and destroyed. The pupae in tunnels be destroyed by introducing iron hook into the tunnel.

For the control of stem borers, the tunnel be filled with pesticides and plugged with mud. The larvae will be killed in the tunnels of the stem. The tunnels also be filled with kerosene or petroleum and the mouth be plugged with the mud. This will give good control of pests.

Curative Control Measures

Cultural Control

1. Destruction of alternative hosts. *e.g., H. puera* is key pest of teak *T. grandis* but can cause damage to *Vitex negundo, Heteropragma* sp., *Millingtonia* sp., etc such alternative plants be removed from target area.
2. Retaining strips and patches of pre-existing mixed trees ex: *Ailanthus* be grown in mixture of sisso, gamhar etc.
3. Pruning the leaves/twigs at peak period of infestation. This is possible on Maharukh (*Ailanthus* sp.) for the pest *Atteva fabriciella.*
4. Promoting early maturing of leaves by management of plantation by irrigation. This is possible in case of *P. reflexa* on sisso tree of forest.
5. Treating the surface with pesticides where insect stages are going to fall down. *e.g., P. reflexa* pest of sisso.

Mechanical Control

1. Use of light traps for attracting moths and later, killing them.
2. Use of iron hooks in tunnel for killing borers inside the tunnel. Locating the tunnel and probing the same with split bamboo.
3. Use of slippery bands of metals or other Sticky materials for avoiding ascending pests on trees.
4. Scrapping bark of the tree trunk with wooden knife for removing pest stages from bank.
5. Pheromone traps be use for catching and destructing moths of forest pests. *e.g., H. robusta.*

Biological Control

The eggs, larvae and pupae of lepidopterous pests are parasitized or predated by parasitoids or predators in field condition. Biocontrol agents should be mass reared and used to control forest pest Lepidoptera.

Parasitoids

Important parasitoids parasiting forest pest Lepidoptera are given in Table 51.

Table 51: Parasitoids of Forest Pest Lepidoptera

Sl.No.	Pest	Host Plant	Parasitoid
1.	*Arthroschista hilaralis*	Kadamba *Anthocephalus kadamba*	*Apanteles balteatea* (Braconidae) *Cedria paradoxa*
2.	*Atteva fabriciella*	Maharukh *Ailanthus excelsa*	*Brachymeria himeatteva* (Chalcidae) *Bessa remota. Carcelia* sp.(Tachinidae)
3.	*Tonica niviferana*	Semul *Bombax ciba*	*Xanthopimpla brevicauda* (Ichneumonidae)
4.	*Hypsipyla robusta*	Toon *Cedrela toona* Mahogany	*Trichogramma robusta* (Trichogrammatidae) *Tetrastichus spirabilis* (Chalcidae) *Antrocephalus destructor*
5.	*Hapalia machaeralis*	Teak *Tectona grandis*	*Apanteles ruidus* *Apanteles machaeralis* (Braconidae)
6.	*Hypsipyla robusta*	Mahogany	*Apanteles hypsipylae* *Cotesia ruficrus* (Braconidae)
7.	*Hypsa alciphron*		*Apanteles priscus* (Braconidae)
8.	*Lymantria* sp.	*Shorea* sp.	*Apanteles cacao* (Braconidae)
9.	*Euproctis lunata*	Babul, khair	*Apanteles lunata* (Braconidae)
10.	*Helicoverpa armigera*	Polyphagus	*C. ruficrus* (Braconidae) *Campoletis chlorideae* (Ichneumonidae)

Predators

Predatory insects of forest Lepidoptera refer to

1. Lady bird beetles — Feed on eggs and early instars of larvae
2. Tiger beetles — Feed on eggs and early instars of larvae
3. Mantids — Feeds on eggs, larvae, pupae and adults of forest Lepidoptera
4. Dragonflies — Feed on micro lepidopteran moths, pyraustids, pyralids, etc.
5. Pentatomid bugs — Caterpillars
6. Reduviid bugs — Caterpillars

Chemical Control

Insecticides give quick results for pest control but they have many side effects such as development of resistance in pests, pest resurgence, secondary pest outbreak, pollutions and health hazards etc. Therefore, they should be used with great care. The application frequencies of pesticides and doses should be increased without consultation of experts or entomologists. List of important insecticides for treating forest trees for control of insect pests is given in Table 52.

Table 52: Pesticidal Doses for some Lepidoptera

Sl.No.	Pest	Insecticides	Dose	
			Dusting	Spraying
1.	*Indarbela tetraonis* and *A. hilaralis*	Carbaryl Endosulfan	–	0.1 to 0.15 per cent 0.03 per cent
2.	*Atteva fabriciella* and other defoliators and leaf miners	Malathion Endosulfan Aldrin	– – 5 per cent	0.2 per cent 0.4 per cent
3.	*T. niviferana* *N. velutina*	Rogor Rogor	– –	0.02 per cent 0.02 per cent
4.	*P. reflexa*	Methyl parathion quinolphos BHC	– 5 per cent	0.03 per cent 0.03 per cent –
5.	*H. puera* *H. machaeralis*	Carbaryl Carbaryl Fenitrothion	5 per cent 5 per cent –	0.15 per cent 0.15 per cent 0.03 per cent
6.	For borers and gall formers	Aldrin Endosulfan	–	0.03 per cent 0.03 per cent
7.	General fumigation	Paradichlorobenzene Ethylene dibromide		

9

Summary and Conclusion

The first chapter deals with general introduction. Which narrets national and international status of the topic.

Second topic deals with review of literature and third is related to collection and preservation of Lepidoptera and methods adopted for present work. Fourth chapter is devoted to biodiversity. Fifth to biology and intrinsic rates of increase and Sixth to checklist of moths found in western Ghats of Maharashtra specially Kolhapur and Satara districts. Seventh chapter is forpest Lepidoptera concepts and eighth for pest Lepidoptera control. In all twenty three species of Lepidoptera, have been described in the text. Out of which 13 were found new to the science and 10 species have been redescribed for fulfilling the gap of additional characters and more valid characters which were useful in identification systems.

The chapter fifth embodies details of biology of three lepidopterous insects specially,

1. *E. narcissus,*

2. *E. machaeralis,* and

3. *H. producta.*

The biology has been studied with respect to life cycle,host plants, nature of damage and morphometry of immature stages, adults stages and seasonal distribution in western Ghats.

Biology of *E. narcissus* reaveled that a single female laid an average of 151.8 ± 8.58 yellowish white eggs on the surface of tender leaves. Preoviposition, oviposition and postoviposition periods lasted for 3.5 ± 71, 5.0± 0.67 and 1.5 ± 0.53 days respectively. Incubation period was 3.42 ± 0.44 days with hatching percentage of 75

± 10. Larva developed in 22.15 ± 0.93 days through 5 instars. The full grown larva pupated on leaves in petridish. Pupal period was 5.75 ± 0.67 days. Adult was medium sized. Longevity of male and female moths were 8.7 ± 85 and 9.8 ± 1.03 days respectively. Biology of teak skeletonizor, *E. machaeralis* revealed that a female laid an average of 187.5 ± 11.5 yellowish white eggs singly on the leaves. Preoviposition, oviposition, and postoviposition period lasted for 2.60 ± 0.52, 4.7± 0.67 and 1.4 ± 0.52 days respectively. Incubation period was 3.25 ± 0.38 days with hatching percentage of 78.5 ± 9.88.The moth was developed within 17.0 ± 1.12 days through 5 instars. A freshly hatched larva measured 0.18 ± 0.011 mm in head width. The prepual period lasted for 1.45 ± 0.51 days. Pupal period was 5.70 ± 0.47 days. Freshly emerged adult was medium sized with bright yellow coloured fore wings and with pinkish transverse zigzag lines. Longevity of male and female moths were 5.10 and 7.0 ± 0.79 days respectively. Male moth measured 11.8 ± 0.51 mm in length, 1.7± 0.98 mm in breadth and with 18.6 ± 0.73 mm wing expanse. Female moth measured 9.04 ± 0.15 mm in length, 2.3 ± 0.073 mm in breadth and with 19.3 ± 0.92 mm wing expanse. The sex ratio for male to female was 1:1.5. Life cycle was completed in 27.40 ± 2.48 days. Biological studies of *H. producta* revealed that a female laid 157.9 ± 6.89 eggs on the leaves of *F. glomerata*. Preoviposition, ovipostion and postoviposition periods lasted for 3.4 ± 0.52, 3.8 ± 0.79 and 2.4 ± 0.52 days respectively. Incubation period was 3.0 ± 0.23 days with hatching percentage of 75 ± 6.07. Larva developed in 16 to 22 (mean 19.1 ± 1.36) days through 5 instars. Pupal period was 6.2 ± 0.41 days. Adult was medium sized. Longevity of male and female moths were 6.35 ± 62 and 10.15 ± 0.97 days respectively.The sex ratio (male : female) was 1:2. Life cycle was completed in 29.5 ± 2.41 days.

Intrinsic rates of increase under laboratory conditions in *E. narcissus, E. machaeralis, H. producta* were 0.129, 0.149 and 0.141 per female per day and population multiplied 75.88 times in mean generation time 'T' 33.54 days, 88.16 times in mean generation time 'T' 30.06 days and 76.76 times in mean generation time 'T' of 30.78 days respectively.

A checklist of moths found in western Ghats of Kolhapur and Satara have been incarporated in the book. In all, 103 species of moths have been indexed. The chapter nineth deals with summary and conclusion. The book is also incorporated with bibliography refered for the completion of present work.

In western Ghats of Kolhapur and Satara districts harmful species of moths are abundant, many Noctuids, Hypsids, Sphingids, Arctiids, Gelechids and Pyralids are destructive at their larval stage and hence their control for protection of forest is extremely essential. Thirteen new species of moths belonging to familes Arctiidae (7), Sphindae (2), Noctuidae (3), Hypsidae (1) have been described in the text while 10 species of moths belonging to the families Arctiidae (2), Sphingidae (3), Noctuidae (3), Hypsidae (1), Pyralidae (1) have been redescribed for fulfilling the gap of additional characters.

The book contains a very special chapter (VIII) for control strategies of lepidopterous pest of forest under which preventive, cultural, mechanical,biological and chemical control are given.

Western Ghats is amongsts 18 hot spots of the world for protection and conservation of biodiversity. From western Ghats of Kolhapur and Satara thirteen new species of moths have been reported, damaging various plants. Biology of three lepidopterous species *viz.*, *E. narcissus, E. machaeralis* and *H. producta* have been studied under laboratory conditions (25±2°C, 65 ± 5 per cent R.H.,12 hr photoperiod) on their natural host plants, *T. grandis, A. excelsa, F. glomerata* respectively. The life cycle of *E. narcissus* was completed in 32.47 ± 2.41 days and fecundity rate was 151.8±8.58 eggs per female while, *E. machaeralis* taken 27.50 ± 2.48 days and fecundity rate was 187.4 ± 11.5 eggs per female and in *H. producta* the life cycle completed within 29.5 ± 2.41 days and fecundity rate was 157.9±6.89 eggs per female. As compared to other species *E. machaeralis* was more fecund and its intrinsic rate of increase was also highest. Intrinsic rates of increase under laboratory conditions in *E. narcissus, E. machaeralis* and *H. producta* were 0.129,0.149 and 0.141 per female per day and population multiplied 75.78 times in mean generation time 'T' 33.54 days, 88.16 times in mean generation 'T' 30.06 days and 76.76 times in mean generation time 'T' of 30.78 days respectively.

In all, 103 species of moths have been reported from western Ghats of Kolhapur and Satara districts of Maharashtra. Out of which 97 species were common and abundant. Six species were rare in the region.

The present book will be helpful for suggesting appropriate control strategies of lepidopterous pest species and protecting economically important trees like *T. grandis, F. glomerata* and *A. excels* and several others.

Bibliography

Aitkenhead, P. C., Baker, R.B. and G.W.D.D.E. Chicker, 1974. New or common plant diseases and pests: An outbreak of *Spodoptera litura* a new pest under glass in Britain, *Pl. Path.,* 23, 117 -118.

Al-Ali, A., Al –Nejamy, I. K., Abbas, S. A. and A.M.E. Abdul-Masih, 1975. Observations on the biology of the potato tuber moth *Phthorimaea operculella* Zeller (Lepidoptera : Gelechiidae) in Iraq. *Z. Angew. Ent.,* 79, 345-351.

Ali, S. 1981. Do you know these vanishing birds? *Hornbill,* 24–28.

Allen, W. A. 1983. Incidence of *Zoophthora phytonomi* (Zygomycetes: Entomophthorales) in *Hypera postica* (Coleoptera: Curculionidae) larvae in Virginia. *Environmental Entomology,* 12(5), 1318-1321(4).

Ananthakrishnan, T. N. 1978. Impact of deforestation and monoculture plantation on insect fauna of western Ghats. *Proc. Seminar on Ecodevelopment of western Ghats,* 1986, pp.238-243.

Ananthakrishnan, T. N. 1993. Biological diversity concepts and conservation strategies. *Hexapoda,* 5, 5-8.

Andrew, E.A. 1915. Noeson insect pests of green manures and shade trees. *Qtrly J. Scient. Dep. Ind. Tea Assoc,* Calcutta, Part III, pp. 57-62.

Annie John and D. Muraleedharan, 1989. Biology and morphometrices of castor semilooper, *Achoea janata* Linn. (Lepidoptera : Noctuidae). *Uttar Pradesh J. Zool., 9(1), 48-55.*

Anonymous, 1982. Flora and Fauna of Silent Valley, Attapadi and Sabarigin Forest study team Report, Govt. of Kerala, Trivendraum. Compiled by B.K.Nayar,Culcutta Uni., 108.

Ayyar and Ayyar, 1938. On the ecological history of the western Ghats. *Abstract,* pp.98.

Babu, M. H., Bhattacharya, A. K. and Y.S. Rathore, 1979. Rate of intake, growth and digestibility of three lepidopterous insects on soyabean and green gram. *Z. Angew Ent.,* 87(3), 322 -327.

Bains, S. B. and K. K. Shukla, 1976. Effect of temperature on the development and survival of maize borer *Chilo pertellus* (Swinhoe). *Indian J. Ecol.,* 3(2), 149 – 155.

Banerjee, T. C. and N. Haque, 1983. Observation on the biology of *Diacrisia casignetum* Kollar (Lepidoptera : Arctiidae) on sunflower. *Indian J. agri. Sci.,* 53, 372-370.

Barbosa, P. and J. Green blatt, 1979. Suitability, digestibility and assimilation of various host plants of gypsy moth *Lymantria dispar. Oecologia,* 43(1), 111-119.

Bariola, L. A. 1978. Suicidal emergence and reproduction by overwintered pink ballworm moths (Lepidoptera: Gelechiidae). *Environ. Ent.,* 7, 189-192.

Barlow, N.D. and M. N. Clout, 1983. A comparison of 3-parameter, single-species population models, in relation to the management of brushtail possums in New Zealand. *Oecologia,* 60, 250-258.

Bassi, A. 1972. Food preferences in the larvae of two moths *Spodoptera litura* F. (Fam. Noctuidae) and *Diacrisia obliqua* Walk (Fam. Arctiidae). *J. Bombay N. Hist. Soc.,* 71 (1), 161 – 163.

Basu, A.C. 1944. Effect of different foods on the larval and post larval development of the moth *Prodenia litura* F. (Lepidoptera : Noctuidae). *J. Bombay Nat. Hist. Soc.,* 44 (1 and 2), 275-288.

Beeson, C.F.C. 1913. Annual report of Forest Entomology. Scientific Advice for India, pp. 8-11.

Beeson, C.F.C. 1939. Report on Forest Research in India Entomological branch 1937 - 38 part I, 33-39.

Beeson, C.F.C. 1941. Ecology and control of forest insects of India and the neighbouring countries. Government of India (1961 Reprint), pp. 767.

Bell, T.R.D. and Scott, 1991. The Fauna of British India including Ceylon and Burma: Moths-V, Sphingidae, DWJ, B.V. Publishers, pp. 1-537.

Berryman, A. A. 1986. Forest Insects : Principles and practice of population management, Plenum Press, New York, pp. 279.

Berryman, A.A. and W. Baltensweiler, 1981. Population dynamics of forest insects and the management of future forests. *Proc. 17th IUFRO Word Congr., Kyoto,* 2, 423-230.

Bhalani, P. A. 1989. Suitability of host plants for growth and development of leaf eating caterpillar *Spodoptera litura* (Fabr.). *Indian J. Ent.,* 51(3), 427-430.

Bhalani, P. A. and G. M. Talati, 1984. Growth and development of *Spodoptera litura* (F.) on certain food plants. *GAU Res. J.,* 9 (2), 60-62.

Bhat, N. S. and A. K. Bhattacharya, 1978. Consumption and utilization of soyabean by *Spodoptera litura* (Fabricius) at different temperatures. *Indian J. Ent.,* 40 (1), 16-25.

Bhoje, P. M. and T. V. Sathe, 2003. Faunistic studies on butterflies from Radhanagari Wild Life Sanctuary, India. *Indian J. Environ. and Ecoplan.,7(3),* 655-658.

Bhowmick, A.K. and S.M. Vaishampayan, 1986. Observations on the activity of teak defoliator *Hyblaea puera* Cram. on teak (*Tectona grandis*) influenced by the movement of monsoon. *J. Trop. For.,* 5(1), 27-35.

Bilapate, G. G., and V.M. Pawar, 1980. Life fecundity tables for *Heliothis armigera* Hubner (Lepidoptera : Noctuidae) on sorghum earhead. *Proc. Indian Acad. Sci.* (Ani. Sci.), 89 (1), 69-73. (W. L. 2555).

Birch, L. C. 1948. The intrinsic rate of natural increase in an insect population. *J. Anim. Ecol.,* 17, 15-26 (W.L. 25559).

Boardman, L. A. 1977. Insectary culture of *Spodoptera litura* (Lepidoptera : Noctuidae). *N. Z. Ent.,* 6, 316-318.

Borkovec, A. B., Wood, C.W. and P. H. Terry, 1978. Bollweevil : chemosterilization by fumigation and dipping. *J. econ. Ent.,* 71, 862-866. British Museum (Nat. Hist.) Zool. Dept. 626 p.

Broodryk, S.W. 1971. Ecological investigations on the potato tuber moth *Phthorimaea opereulella* Zeller (Lepidoptera : Gelechiidae). *Phytophylactica,* 3, 73-84.

Butani, D. K. 1979. *Insects and Fruits,* Periodical Export book agency D-42, Vivek Vihar, Delhi, 415 pp.

Butt, B.A. and E. Cantu, 1962. Sex determination of Lepidoterous pupa. U. S. Department of *Agriculture Pamphlet ARS,* 7, 33-75.

Champion, H. G. and S. K. Seth, 1968. A revised survey of the forest type of India. Manager of Publications, Delhi, pp. 404.

Champion, H.G. 1934. The effect of defoliation on the increment of teak saplings. *For. Bull.* (Silv.), No. 89.

Chandrasakhara, C. S. 1984. Western Ghats Development Programme : Approach to the Seventh Five Year Plan 1985-90, pp. 10-19. *Hexapoda, 5,* 17-53.

Chatterjee, M. and D. K. Chaudhuri, 1989. On the consumption and utilization of castor leaves by the larvae of *Diacrisia casignetum* Kollar (Lepidoptera: Arctiidae). *Indian Biologist,* 221, 23-27.

Chatterjee, S. N. 1967. The identity of *Spodoptera mauritia acronyctoides* Guenée, *Spodoptera pectin* Guenée and *Spodoptera abyssinia* Guenée (Lepidoptera:Noctuidae) based on a comparative study of the male and female genitalia. *Nat. Inst. Sci. India Proc.*, 35B(1), 46-52.

Chelliah, S. and K. Srinivasan, 1987. Biology and management of diamondback moth in India. *In diamondback moth management, Proceedings of the International Workshop Tainan,* Taiwan.

Choudhary, R. R. P. and A. K. Bhattacharya, 1986. Bio-ecology of lepidopterous insects on winged bean *Psophocarpus tetragenolabus* (L.) D.C. Memoir No.11, Entomological Society of India, New Delhi, p. 130.

Chu, Y. I., Wang, S.C. and S. H. Cheug, 1976. Studies on the mass production of the tobacco cutworm *Spodoptera litura* (Fabricius). *Plant Protection Bulletin, Taiwan,* 18, 173-182.

Crumb, S. E. 1956. The larvae of the Phalaenides. *Tech. Bull.* USDA, 1135, 1-365.

Da Costa, M.A. and S.J. Weller, 2005. Phylogeny and classification of Callimorphini (Lepidoptera: Arctiidae: Arctiinae). *Zootaxa,* 1025, 1–94.

Daniels, J. R. 2000. Animal species diversity in the western Ghats : CES Technical report No. 5.

David, B.V. and T. Kumarswami, 1982. Insect pests of forest trees. Elements of Economic Entomology. Popular Book Depot, Madras pp. 177- 187.

Deshmukh, P. O., Rathore, Y. S. and A. K. Bhattacharya, 1976. Host range of Bihar hairy catterpiller *Diacrisia obliqua* Walker. *Bull. Ent.,* 17, 85-99.

Doresesta, S. E. and M. Nieves, 1968. Laboratory studies on the life cycle of the tobacco, potato and tomato leafminer *Phthorimaea operculella. Agron. Trop. (Maracay),* 18, 461-474.

Dreze, J. and A.K. Sen, 1995a. India : The political economy and Hunger : Selected essays. Clareudon press. (abdridgded).

Dreze, J. and A. Sen, 1995b. India : Economic development and social opportunity. New Delhi : Oxford University Press, pp. 291-292.

Dubatolov, V.V. 2006. Cladogenesis of tiger-moths of the subfamily Arctiinae, development of a cladogenetic model of the tribe Callimorphini (Lepidoptera : Arctiidae) by the SYNAP method. *Euroasian Entomological Journal,* 5(2), 95–104.

Dubatolov, V.V. 2008. Construction of the phylogenetic model for the genera of the tribe Arctiinae (Lepidoptera : Arctiidae) with the SYNAP method. *Entomological Review,* 88(7), 833-837. Translated from: *Entomologicheskoe Obozrenie,* 87(3), 653–658.

Dubatolov, V.V. 2010. Tiger-moths of Eurasia (Lepidoptera : Arctiidae) (Nyctemerini by Rob de Vos and Vladimir V. Dubatolov). *Neue Entomologische Nachrichten,* 65, 1–106

Dumbre, R. B., Khanvilkar V. G., Dalvi C.S. and S. R. Bhole, 1989. Field studies on tasar silkworm rearing in the Konkan Region. *Indian J. Ent.,* 51(1), 76-83.

Duponchel, P.A.J. 1844-1846. Catalogue methodigue das Lepidopteres d' Europe. XXX + 425pp, Paris.

Edwards, E. D. 1996. Arctiidae In : Nielsen, E.S., Edwards, E.D. and T.V. Rangsi (eds.), *Checklist of the Lepidoptera of Australia, Monographs on Australian Lepidoptera.* CSIRO Clayton, 278-286.

Evans, F.C. and F.E. Smith, 1952. The intrinsic rate of natural increase for the human louse, *Pediculus humanus* L. *Am. Nat.*, 86, 299-310.

Fabricius, J. C. 1793. Entomologia systematica emendata et. aucta- Hafniae, Impensis, C. G. *Profit, fil. et. Soc.,* 3(2), 350.

Feng, H.T. 1987. Control economics of resistant diamondback moth-a field case study. *Plant Protection Bulletin Taiwan,* 29(2), 175-84.

Ferguson, D. C. and P.A. Opler, 2006. Checklist of the Arctiidae (Lepidoptera: Insecta) of the continental United States and Canada. *Zootaxa,* 1299: 1–33.

Fletcher, T. 1914. Some South Indian insects and other animals of importance, M/s. Bishen Singh Mahendra Pal Singh, 29.A, New Connaught Place. Dehra Dun. pp. 448-443.

Fletcher, T. B. 1917. Report of Proccedings of the second Entomological Meeting, Pusa, 1917.

Foot, M. A. 1979. Bionomics of the potato tuber moth, *Phthorimaea operculella* (Lepidoptera : Gelechiidae) at Pukehohe. *N.Z.J. Zool.,* 6, 623-636.

Forster, W. and T. A. Wohlfahrt, 1956-71. Die Schmetterlinge Mitteleuropas 3 and 4, Eulen (Noctuidae), vii: Stuttgart pp. 329.

Gadekar, R. R., Chandrasekhara, K. and P. Nair, 1990. Insect species diversity in the tropics: sampling methods and a case study. *J. Bomb. Nat. Hist. Soc.,* 87, 353-357.

Gaonkar, H. 1996. The butterflies of Western Ghats, India and Shrilanka. A biodiversity assessment on the ecological history of western Ghats. Abst., pp.90.

Garthwaite, D. F. 1939. Biology of *Calopepla leayana* Latr. (Chrysomelidae : Coleoptera) and the possibilities of control. *Indian For. Rec.,* 5(2), 237-241.

Gaur, J.P. 1992. Races of *Antheraea mylitta,* the trophical tasar silkmoth, their distribution and variability. *Indian J. Ent.,* 54(3), 275-284.

Ghorpade, B. R. and S. P. Patil, 1991. Insect pest recorded on forest trees in the Kokan region of Maharashtra state (India). *Indian J. For.,* 14(3), 245-246.

Gomma, A. A. S., Et.-Sherif, Salem, A.A. and I.A. Hemeida, 1979. On the biology of potato tuber worm, *Phthorimaea operculella* Zeller (Lepidoptera:Gelechiidae), Reaction of Photoperiodism. *Z. Angew. Ent.,* 87, 430-435.

Goyal, S.P. and V. S. Rathore, 1988. Pattern of insect plant relationship determining susceptibility of different hosts to *Heliothis armigera* Hubner. *Indian J. Ent.,* 50(2), 193-201.

Grosscurt, A. C. 1978. Diflubenzuron, some aspect of its ovicidal and larvicidal mode of action and an evaluation of its practical possibilities. *Pest. Sci.,* 9, 372-386.

Grote, A. R. 1890. North American Lepidoptera. Revised check-list of the North American Noctuidae, part I, Thyatirinae, Noctuinae, Bremen pp. 52.

Grote, A. R. 1882. *New Checklist of North American Moths,* New York pp. 73.

Gubbaiah and T. S. Thontadarya, 1977. Bionomics of the potato tuber worm, *Gnorimoschema operculella* Zeller (Lepidoptera : Gelechiidae) in Karnataka. *Mysore J. agric. Sci.*, 2, 380-386.

Guene'e, A. 1841. Essai sur la classification des Noctuelides (suite) (1). *Annales de la Soc Entomologique de* France, 10, 53-83.

Guene'e, A. 1852-54. In Boisduval, J. A. and A. Guenee, Histoire naturelle des Insects species General des. Lepidopteres. 5 Noctuelides, I. xcvi + 407, pp. 6, Noctuelites, II, 444, pp. 7. Noctuelites, 111, 442, pp. 8. Deltoides et Pyralites, pp. 448, Paris.

Guene'e, A. 1837. Essai pour servir a la classification dea Noctuelides (suite). *Annales de la Soc. Entomologique de* France, 6, 311-367.

Gupta, M and A. D. Pawar, 1985. Multiplications of *Telenomus remus* Nixon on *Spodoptera litura* (Fabricius) reared on artificial diet. *J. Adv. Zool.*, 6, 13-17.

Hampson, G. F. 1892. The Fauna of British India, including, Ceylon and Burma, Moth 1 : VIII pp. 527, London.

Hampson, G. F. 1895. The fauna of British India, including Ceylon and Burma, Moth 3 : XXXVIII + 546 pp., London.

Hampson, G. F. 1897. The moths of India. Supplementary paper to the volume in "The Fauna of British India", Part I. *J. Bombay Nat. Hist. Soc.*, 11, 277-297, Part II, t.c. 438-462, pi. A (Lepidoptera).

Hampson, G. F. 1902. The moths of South Africa (Part II). *Ann. South Afri. Museum*, 2, 255-446.

Hampson, G. F. 1902-1913. The Lepidoptera-phalanae of the Bahemas. *Ann. Nat. Hist.*, 14, 165-188.

Hampson, G. F. 1908. The moths of South Africa, Part III, Ann. S. Afr. Mus. Cape Town, 3, 389-438.

Hampson, G. F. 1909. Catalogue of The Lepidoptera Phalenj; In The British Museum. v. 8, illus. London. (27) Hart, C. A. 1919. The pentatomoidea of illinois with keys to the nearctic genera. *Ill. Nat. Hist*.

Hampson, G.F. 1910. Zoological collections from Northern Rhodesia and adjacent territories: (Lepidoptera :Phalaenae). *Proceedings of the Zoological Society of London*, 388–510.

Hampson, G. F. 1912. The moths of India, supplementary paper to the volume in "The Fauna of British India", Series IV, Parts HI-V. *J. Bombay Nat. Hist. Soc.*, 21, (411-446, 878-911,1222-1272), 1 Pl (Lepidoptera).

Hampson, G. F. 1913. Catalogue of the Lepidoptera Phalaenae in the British Museum. Vol. 13, 609 p., planches 222-239. Taylor and Francis. London.

Hampson, G. F. 1976. The Fauna of British India, including Ceylon and Burma, Moth 2 : XXXII + pp. 609, London.

Herrich-Schaffer, G. A. W. 1845. Systematische Bearbeitung der schmetterlinge von Europa, Zugleich als Text. Revision und supplement Zu Jakob Flubner's sammlung ouropaischer schmetterlinge.1. Die Schwarmer, Spinner arid Eulen (Hepialides-cossides–zygaenides-sesiides-sphingides Bombycides-Noctuides-Nycteolides), pp. 450, Regensburg.

Holloway, J. D. 1988. The Moths of Borneo 6: Family Arctiidae.

Hough, J. A. and D. Pimental, 1978. Influence of host foliage on development, survival and fecundity of gypsy moth. *Environ. Entomol.,* 7(1), 97-102.

House, H. C. 1963. Nutritional diseases. In E.A. Steinhaus, (ed), *Insect Pathology,* New York, Academia Press, 1, 133-160.

Howe, R. W. 1953. The rapid determination of intrinsic rate of increase of an insect population. *Appl. Biol.,* 40, 134-155.

Howe, R. W.1952, Miscellaneous experiment's with grain weevils. *Entomol. Mon. Mag.,* 88, 252.

Imms, A. D. 1957. A text book of entomology. Chapman and Hall (London).

Inoue, H. and S. Sugi, 1958-1961. *Checklist of the Lepidoptera of Japan 5 Noctuidae and 6 Hyblacidae,* pp. 431-683, Tokyo.

Javed Iqbal Siddiqi, 1985. Studies on Reproduction. III. Mating in *Diacrisia obliqua* Walker (Lepidoptera: Arctiidae). *Indian J. Ent.,* 47(4), 405-409.

Joseph, K. J. 1984. Insect life and ecodevelopment of the western Ghats. *Proc. Seminar on Eco development of western Ghats,* 1986, pp. 84-89.

Joseph, K. J., Narendran, T.C. and M. A. Hag, 1983.Outbreak of hairy caterpillers (Eupterote spp) as serious pest of cardamon in the Mackimalai area of South India and recommendations for their integrated management. *Tropical Management,* 29, 166-172.

Joseph, T.M. 2004. Biodiversity conservation in the western Ghats. *Biodiversity and Environment,* 17, 221-227.

Joseph, T.M. and G.K. Karnavar, 1991. Identification of sex in the larval and pupal stages of Ailanthus defoliator, *Eligma narcissus indica* Roth. (Lepidoptera : Noctuidae). *Entomon,* 14(4), 331 – 333.

Kalia, S. and N.S.K. Harsh, 2003. *Metarhizium anisopliae* (Metschuikoff) sorokin pathogenic to the larva of teak defoliator *Hyblea puera* Cramer. *J. Ent. Res.,* 27(2), 1-2.

Kansu, I. 1962. The effect of the food on the larvae of butterflies and moths, an experiment on the gypsy moth larvae (in Turkish). Ankara Univ. Ziraat Fak. *Yamin,* 2, 116-138.

Kehimkar, I. D. 1997. An introduction to moths NCSTS. *Hornbill Series,* Mumbai- Dec. 1999, pp. 379.

Khan, H. R., Kumar, S. and L. Prasad, 1988. Studies on seasonal activity of some Agroforestry insect pests by light lamp. *Indian J. For.,* 114(4), 215-229.

Khan, M. M. and B.B.L. Srivastav, 1993. Biological interaction of penfluron against *Euproctis icilia* Stoll. (Lepidoptera : Lymantriidae). *Indian J. Ent.* 55(3), 267-274.

Khoshoo, T. N. 1995. Census of India's Biodiversity tasks ahead. *Curr. Sci.,* 69, 14 -17.

Khoshoo, T. N. 1996. India needs a National Conservation Board. *Curr. Sci.* 71(7), 506-513.

Kitching, I. J. 1984. An histological review of the higher classification of the Noctuidae (Lepidoptera). *Bull. Br. Mus. Nat. Hist. (Ent.),* 49(3), 153-234.

Kitching, I. J. and J.M. Cadiou, 2000. *Hawkmoths of the world; annotated and illustrated revisionary checklist (Lepidoptera : Sphingidae),* X of 227 pp., 8 pis. Ithaca and London, USA and UK; Cornell University Press.

Knipling, E.F. 1966. Introduction in insect colonization and mass production (Smith, C.N. ed.), Academic Press, New York pp. 1-2.

Kundu, B. C. 1956. A review of research works carried out at the Jute Agricultural Research Institute 1948-49 to 1955-56, Indian Central Jute Committee, pp. 109-122.

Lall, B. S. 1949. Preliminary observations on the bionomics of potato tuber moth, *Gnorimoschema operculella* Zeller and its control in Bihar. *Indian J. Agric. Sci.,* 19, 295-305.

Latch, G.C.M. 1965. *Metarhizium anisoliae* (Metsch), Sorok Strains in New Zealand and the possible use for controlling pastur inhabiting insects. *NZ. J. Ag Res.,* 8, 384-396.

Layne, J. R. Jr. and D. K. Kuharsky, 2000. Trigering of cryoprotectant synthesis, in the wooly bear caterpillar *Pynharctia isabella* (Lepidoptera : Arctiidae). *J. Exp. Zool.,* 286A, 367-371.

Le Roux, E. J. 1963. population dynamics of agricultural and forest insect pests. *Mem. Ent. Soc. Can.,* 32,1-103.

Lefroy, H. M. 1907. The more important insects injurious to Indian Agriculture. Mem. Dept. Agric. India, 1, 160.

Lefroy, H. M. 1906. *Indian Insect Pests,* Calcutta, p. 151.

Lefroy, H. M. 1909. Indian Insect Life, Calcutta and Simla, Thaeka, Spin and Co.

Linnaeus, C. 1758. System Naturae Edn. 10, Regnum Ahimale, Holmaie.1, 824.

Liu, H., Chi, H. Chen, C.N. and K.S. Kung, 1987. Population parameter of the diamondback moth in relation to its susceptibility to insecticides. *Plant Protection Bulletin* Taiwan, 29(3), 283-291.

Lotka, A. J. 1925. Elements of Physical Biology. Williams and Wilkins, Baltmore Md – pp. 462.

Madhuchanda, Chatterjee and D. K. Chaudhuri 1992. Food Selection by the Arctiid Moth *Diacrisia casignetum* Kollar. *Indian J. Ent.,* 54(4), 174-180.

Malik, R. A. Punjabi, A. A. and A.A. Bhat, 1972. Survey and study of insect and non-insect present in Kashmir. *Horti.,* 3, 29-44.

Mani, M.S. 1993. Modern Classification of Insects. Pub. Satish Book Enterprise, Moti Katra, Agra, India pp. 1-331.

Masoodi, M. A. 1985. Growth response of *Lymantria obfuscata* Walker in relation to tannin content in different host foliages. *Indian J. Ent.,* 47(4), 422-426.

Masoodi, M. A. and A. S. Srivastava, 1985. Effect of host plants on the pupal weight and fecundity of *Lymantria obfuscata* Walker (Lymantridae : Lepidoptera). *Indian J. Ent.,* 47(4), 410-442.

Mathew, G. and M.J.M. Ali, 1987. Microbial pathogens causing mortality in the carpenterworm *Cossus cadambae* Moore (Lepidoptera : Cossidae), A pest of teak (*Tectona grandis* Lin.) in Kerala (India). *J. Trop. For.,* 3(4), 349-351.

Mathew, G. and K.S.S. Nair, 1984. Bagworms (Lepidoptera: Psychidae) of Kerala. Their potential as pests of tree crops. *Proc. 3ʳᵈ Oriental Entomology Symp.* Trivandrum, 2, 163-167.

Mathew, G. 1982. A survey of beetles damaging commercially important stored timber in Kerala KFR Research Report 10, Kerala Forest Research Institute, Peechi, pp. 92.

Mathew, G. 1990. An artificial diet for the Teak defoliator, *Hyblea puera* Cramer (Lepidoptera : Hyblaeidae). *Entomon,* 15 (3), 159 -163.

Mathew, G. and M. G. R. Menon, 1986 c. Identification of some Indian Pyraustinae (Lepidoptera : Pyraustidae). *Journal of Entomological Research.*

Mathew, G. and K. S. S. Nair, 1985. Insects associated with forest plantations of *Parascrienthes falcataria* in Kerala, India. *Malay, Forester,* 48(3), 200-205.

Mathew, G. and M.G.R. Menon, 1986 a. Identification of some leaf rollers belonging to the genera *Bradina, Marasmia and Cuaphalocrocis* (Lepidoptera : Pyraustidae). *Entomon.*

Mathew, G. and M.G.R. Menon, 1986 b. Genitalial morphology of some Indian *Nymphulinae* (Lepidoptera: Pyraustidae) *Journal of Entomological Research.*

Mathew, G. 1986. Insects associated with forest plantations of *Gmelina arborea* in Kerala, India. *Indian J. For.,* 9 (4), 308-311.

Mathur, R. N. 1941. Catalogue of food plants and their noctuid defollators. *Indian For. Rec.,* 7, 143 – 151.

Mathur, R. N. and B. Singh, 1960. A list of insect pests of forest plants in India and the adjacent countries for the use of forest officers, part 6. *Indian For. Bull. (Ent.),* (7), 53-54.

Mathur, R. N. and B. Singh, 1961. A list of insect pests of forest plants in India and the adjacent countries for the use of forest officers, Part 9. *Indian For. Bull* (Ent.), 171(9), 28-38.

Mcleod, J.M. 1972. A comparision of discrimination and of density responses during oviposition by *Exenterus amictorins* and *E. depronis* (Hymenoptera : Ichneumonidae), parasites of *Neodiprion swinei* (Hymenoptera : Diprionidae). *Can. Entomol.,* 109, 789-796.

Merker, E. 1964. The effect of food on the development of *Lymantria dispar* L. (in German). *Allg. Forst. Jagdztg,* 135 (2), 3436.

Morris, R. F. and C. S. Miller, 1954. The developments of life tables for the spruce budworm. *Can. J. Zool.,* 32, 382-301.

Mosher, E. 1961. A classification on the Lepidoptera based on characters of the pupa. *Bull. 111. State Lab. Nat. Hist.,* 12, 14-159.

Mukharjee, A. K. 1949. Life history and bionomics of the potato tuber moth at Allahabad together with some notes on the external morphology of the immature stages. *J. Zool. Soc. India,* 1, 57-67.

Munroe, E. 1960. New tropical Pyraustinae (Lepidoptera : Pyralidae). *Can. Ent.,* XCII(3), 164-173.

Munroe, E. 1969. Contribution to a study of the Pyraustinae (Lepidoptera : Pyralidae) of Temperate East Asia. VII. *Can. Ent.,* 101, 1069-1077.

Munroe, E. 1970. Contribution to a study of the Pyraustinae (Lepidoptera : Pyralidae) of Temperate East Asia. IX. *Can. Ent.,* 102(3).

Nagarkatti, S. and S. Prakash, 1974. Rearing *Heliothes armigera* (Hubn.) on an artificial diet. *Tech. Bull.* No. 17-CIBC, Banglore 169-173.

Nair, K.S.S. and G. Mathew, 1992. Biology, Infestation Characteristics and impact of the bagworm *Pteroma plagiophleis* Hamps. in forest plantation of *Paraserianthes falcafaria. Entomon,* 17 (1 and 2), 1-13.

Nair, K.S.S. 1982. Seasonal incidence, host range and control of teak sapling borer, *Sahyadressus malabarious,* K. F. R. I. Ras, Rap. 16-48.

Nair, K.S.S. and G. Mathew, 1997. Search for teak trees resistant to the defoliator, *Hyblaea puera* Cramer (Lepidoptera: Hyblaeidae) In Ed. Raman. A. *Ecology and Evaluation of plant feeding insects in natural and man made environment,* 109-122.

Nair, K.S.S., Mathew, G. and M. Sivarajan, 1981. Occurrence of the bagworm *Pterowea plagiophleps,* Hampson (Lepidoptera: Sychidae). as a pest of the tree, *Albizia falcataria* in Kerala, India. *Entomon,* 6(2), 17-180.

Nair, N. C. 1984. Conservation of botanical resources of the western Ghats from a taxonomical points of view. Paper presented in workshop on Ecodevelopment of western Ghats, Trivendrum, 11-13 May, 1984.

Nairnder, Jit Kaur, 1988. Taxonomy of Sub families Catocalinae and Acronictinae (Noctuidae : Lepidoptera) of North and North Eastern India with special reference to external genitalia. Ph.D. Thesis, Punjabi Univ. Patiyala. pp. 307-400.

Narayanan, K., Ramamurthy, V. V., Govindrajan, S. and S. Jayaraj, 1977. Sexing the pupae of gram caterpillar, *Heliothis armigera* Hbn. (Lepidoptera : Noctuidue) in relation to certain morphometric character. *Curr. Sci.,* 46, 192-193.

Nataviria, D. and R.C. Tarumingkeng, 1971. Some important pests of forest trees in Indonesia. *Rimba Indonesia,* 16, 151-165.

Navarajan, Paul, A. V., Das R. and B. Prasad, 1979. Sex determination of pupae of *Heliothis armigera* Hubner, on gram. *Indian J. Ent.*, 41, 285-286.

Nayar, M. P. 1980-82. Endemic Flora of Penninsular India and its significance. *Bull. Bot. Surv. India*, 22,12-23.

Negi, S.S. 1986. A handbook of Forestry International Book Distributors, Dehara Dun pp. 6-7 and 548-554.

Nirula, K. K.1960. Control of potato tuber moth. *Indian Potato J.,* 2, 47-51.

Nye, I. W. B. 1975. The generic names of moths of the world- I, Noctuidae (Part) :Noctuidae, Agaristidae and Nolidae, pp. 568, Trustees of the British Museum (Natural History), London.

Okada, M. 1977. Stuides on the utilization and mass production of nuclear polyhedrosis virus for control of the tobacco cutworm, *Spodoptera litura* (Fabricius). *Rev. Plant. Protection Res.*, 19, 102-1228.

Pachauri, R. K. and P. V. Shridharan, 1998. Green India -2047, Looking back to Think ahead. *Green India,* 31,1-346.

Packard, A. S. 1869. The characters of the lepidopterus family Noctuidae. *Proc. Portland Soc. Nat. Hist.,* 1,153-156.

Painter, R. H. 1936. The food of insects and its relation to resistnace to insect attack. *Am. Nat.,* 708, 547-566.

Painter, R. H. 1968. *Insect Resistnace in Crop Plants.* pp xi + 520 pp.

Pajni, H. R. and H. S. Rose, 1977. Male genitalia of family Pyraustidae (Lepidoptera : Pyraloidea). *Res. Bull. (Sci.),* Panjab Univ. 28 (3 and 4), 131-141.

Pandey, N. D., Yadav, D. R. and T.P.S. Teotia, 1968. Effect of different food plants on larval and post larval development of *D. obliqua* (Walker). *Indian J. Ent.,* 30, 221 - 234.

Pandharbale, A. R. and T. V. Sathe, 2001. On a new species of the genus *Syntomis* (Syntomidae : Lepidoptera) from the environment of western Ghats (Satara district). *Indian J. Environ. and Ecoplan.,* 5(3), 601-602.

Pandharbale, A. R. and T. V. Sathe, 2002. Abundance and distribution of the moths: Families Noctuidae, Sphingidae and Syntomidae from western Ghats and plain region of the Satara district (Maharashtra). *International Conference of SAARC countries on Biotechnology in Agriculture, Industry and Environment,* 2002, Karad, India, Abst.

Pant, N.C. 1973. Nutrition of phytophagous insects, in insect Physiology and Anatomy 2nd ed. (Pant. N.C. and Ghai. S. eds.) pp. 232-238.

Pant, N. C. 1964. Entomology in India silver jubilee number of *Indian Journal of Entomology* (eds.), published by Entomological Society of India, New Delhi, India. pp. 529.

Patel, I. S., Shah, A. H. and N. B. Rote, 1987. Effect of different food plants on development of leaf eating caterpillar, *Spodoptera litura* (F.). *GAU Res. J. R.,* 2, 57- 58.

Patel, R.C., Patel, T.C. and J. K. Patel, 1973. Biology and mass breeding of tobacco caterpillar *Spodoptera litura* (F.). *Israel J. Ent.,* 8, 131 -141.

Patil, B. V. and T.S. Thontadarya, 1983. Seasonal incidence of teak skeletonizer, *Pyrausta machaeralis* Wlk. (Lepidoptera: pyraustidae) in Prabhunagar forest. *Indian J. Ecol.,* 10, 204-209.

Patil, B. V. and T.S.Thontadarya. 1987. Biology of the teak skeletonizer, *Pyrausta machaeralis* Walkar. (Lepidoptera : Pyralidae). *Mysore J. Agric. Sci.,* 21 (1), 32-39.

Pearce, K. G. and S. Hanapi, 1984. *Acherontia lanchesis* a new pest of teak, (*Tectona grandis*) in Malaysia. *Malaysian forester,* 47 (1/2), 80-81.

Pemm, K. H. 1993. New owlet moth species (Lepidoptera : Noctuidae) in USSR. *Entomol. Obozav.,* 62(3), 596-600.

Prem Sagar, Jindla, L.N., Mehta, S. K. and G. S. Mann, 1990. Field Screening of Different Jute Cultivars Against Bihar-hairy Caterpillar *Spilosoma obliqua* Walker in the Punjab. *Indian J. Ent.,* 52(3), 442-444.

Premkumar, M., Dale, D. and M. R. G.K. Nair, 1977. Consumption, digestion and utilization of food by larvae of *Spodoptera litura* F. (Noctuidae : Lepidoptera). *Entomon,* 2, 7 – 10.

Qureshi, Q. G., Shrivastava, R.C., Bhargava, M.C. and G. K. Sharma, 1986. Sex determination at pupal stage in the red hairy catterpillar, *Amsacta moorei* Butler. *Entomon,* 11(2), 157-159.

Raj, B. T. 1988. Seasonal variation in the male population of potato tubermoth *Phthorimaea operculella* Zell. in the Deccan plateau. *Indian J. Ent.,* 50(1), 24-27.

Ramdev, Y.P. and P. J. Rao, 1979 a. Determination of true nitrogen from insect material and insect faeces. *Indian J. Ent.,* 41, 94-96.

Ramdev, Y.P. and P. J. Rao, 1980. Effect of sublethal dose of insecticides on consumptions and utilization of dry matter and diatary constituents of castor, *Ricinus communis* Linn. by the castor semilooper *Achaea janata* Linn. *Indian J. Ent.,* 42, 567-575.

Ramdev, Y.P. and P.J. Rao, 1979 a. Consumption and utilization of dietary constituents of castor, *Ricinus communis* Linn. by the castor semilooper, *Achaea janata* Linn. *Indian J. exp. Biol.,* 17,1154-1157.

Ramdev, Y.P. and P. J. Rao, 1979b. Consumption and utilization of castor. *Ricinus communis* Linn. by Castor semilooper, *Achaea janata* Linn. *Indian J. Ent.,* 41, 260-266.

Ratan, Lal and G. N. Nayak, 1963. Effect of host plants on the development of caterpillar of *Prodenia litura* F. and susceptibility to different insecticides. *Indian J. Ent.,* 25(4), 229 – 306.

Reddy, D.N.R., Kotikal, Y.A. and Kuberappa, 1988. Development and reproduction of *Dasychira mendosa* (Lepidoptera : Lymantridae) on three species of Terminalis. *Indian J. For.,* 11(2), 148 -149.

Reddy, C.G. and Bhattacharya, 1988. Life tables of *Heliothis armigera* (Hubner) on semisynthetic diets. *Indian J. Ent.*, 50 (3), 357-370.

Richards and Waloff, 1961. A study of a natural populaton of *Phytodecta olivacea* Forster (Coleoptera : Chrysomeloidae). *Phil. Trans. R. Soc. Lond.*, 244B, 205 – 257.

Roychoudhary, N., Paul, D.C. and G. Subba Rao, 1991. Growth fecundity and hatchability of eggs of *Bombyx mori* L. in relation to rearing space. *Entomon*, 16 (3), 203-207.

Sagar, P. 1990. Field screening of different jute cultivars against Bihar hairy caterpillar *Spilosoma obliqua* Walker in the Punjab. *Indian J. Ent.*, 52(3), 442-444.

Samarjit, R., Shrivastava, K.M., Saxena, J. D. and S. R. Sinha, 1992. Distribution Pattern of Diamondback moth (*Plutella xylostella* L.) on Cabbage and Cauliflower. *Indian J. Ent.*, 54(3), (262-265).

Sandhu, G.S., Khangura, J.S. and I.S. Deol, 1977. Varietal susceptibility of guava cultivators to bark eating carepillar. *Punjab Hort. J.*, 17 (1 and 2), 62-63.

Sanjayan, K. P. and K. Muragan, 1987. Nutritional influence on the growth and reproduction of two species of Acridies (Orthoptera : Insecta). *Proc. Indian Acad. Sci.*, 96, 229-237.

Santosh Badu, P. D. and V.K.K. Prabhu, 1987. Sexing larva and pupa of *Opisina arenosella* Walker (Lepidoptera : Crytophasidae). *Curr. Sci.*, 14, 735-736.

Sathe, T.V. and A. R. Pandarbale, 2008. Forest pest Lepidoptera. Daya Publishing House, New Delhi, pp. 1-147.

Sathe, T. V. and A. R. Pandharbale, 2004. Biodiversity of moths from western Ghats of Satara district, Maharashtra. Int. Nat. Workshop on Recent Trends in Environmental Sciences, New Delhi. Abts.

Sathe, T. V. 2009. Textbook of Forest Entomology. Daya Publishing House, New Delhi, pp. 1-217.

Sathe, T. V. and A. R. Pandharbale, 1999. Hawk moth (Sphingidae : Lepidoptera) diversity in western Maharashtra including Ghats. *Geobios*, 77-82.

Sathe, T. V. and A. R. Pandharbale, 2001. On a new species of the genus *Syntomis* Hampson (Lepidoptera : Syntomidae). 71st Annual Session, The National Academy Sciences, India, Pune, October, 5 to 7, 2001, Abst. 20, pp. 30.

Sathe, T.V., Inamdar, S. A. and M. V. Santakumar, 1986-87. Fauna of Butterflies from western Maharashtra and western Ghats (Part of Maharashtra only), India./. *Shivaji Uni.* (Science), 23, 391-398.

Sathe, T.V., Mulla, M.K. and D.B. Sathe, 1997. A brief note on the silkworm diversity in western Maharashtra *Proc. 8ᵗʰ Mani. Sci. Congr. Biodiversity and resource Management*, p. 7-8.

Satish, P. M. 1996. Moths and Butterflies of Bhadra Project. M. Sc. Thesis submitted to Kuvempu University, Shimoga, pp. 80.

Saxena, K. N. 1969. Patterns of insect-plant relationship determing susceptibility or resistnace of different plants to an insect. *Int. exp. and appl.*, 12, 751 -766.

Schoonheven, L. M. 1968. Chemosensory basis of host plant selection. *A. Rev. Ent.*, 13, 115-1363.

Sen, A. C. 1954. Keep the potato tuber moth out. *Indian Fing*, 4, 9-10.

Sensarma, P.K. and M.L. Thakur, 1988. Insect factor in the management of forest resources. *My Forest*, 24 (2), 99-113.

Sevastopulo, D. G. 1956. Notes on the Heterocera of Calcutta (Lepidoptera). *Journal of the Bombay Natural History Society*, 54, 302- 308.

Sharma, B. and J. S. Tara, 1985. Insect pests of mulberry plant (*Morus* sp.) in Jammu region of J&K. State. *Indian J. Seri.*, 24, 7-11.

Sharma, B. and J. S. Tara, 1988. Comparison of consumption and utilization of mulberry leaves in two moths *Spodoptera litura* (F.) and *Diacrisia obliqua* walker. *Indian J. Ent.*, 50(3), 336-342.

Sheikh, A. G. 1975. The effect of repeated defoliation caused by *Lymantria abfuscata* Walker on apple tree in Kashmir. *Indian J. Plant Protection*, 3(2), 170-172.

Shorey, H. H. and R. L. Hale, 1965. Mass rearing of larvae of nine noctuid species on simple artificial medium. *J. econ. Ent.*, 58, 522-524.

Shrivastava, S.C. and P.N.Pandey, 1987. Establishment of *Diacrisia obliqua* Walker on certain plants of Compositae. *Indian J. Ent.*, 49(1), 7-15.

Shull, E.M. and N. T. Nadkerny, 1961. Collecting moths by a mercury vapour lamp in Surat Dangs, Gujrat State. *Journal of the Bombay Natural History Society*, 61(2), 281.

Siddigi, J. I. 1982. Observations on the pheromones and physiology of reproduction of certain pests. Doctoral thesis, Aligarh Muslim University, Aligarh.

Siddigi, J.I. 1985. Observations on the secretion of a sex pheromone in *Diacrisia obliqua* Walker (Lepidoptera : Arctiidae). *Zool. Jahrbuch. Physiol.*, 89, 81-87.

Simmons, R. B. and S.E. Weller, 2002. What kind of signals do mimetic tiger moths send ? A phylogenetic test of wasp mimicry systems (Lepidoptera: Arctiidae: Euchromiini). *Proc. Roy. Soc. Lond.*, 269*B*, 983–990

Singh, J. 1979. A taxonomic account of family Noctuidae (Lepidoptera : Noctuidae) of Chandigarh and surrounding areas alongwith the external morphology of *Plusia nigrisigna* Walker, M.Sc. Thesis, Zoology Depatment, P.U., Chandigarh, 1-245.

Singh, S.P. and A.K. Bhattacharya, 1994. Formulation of semisynthetic diets for Bihar hairy caterpillar *Spilosoma obliqua* Walker. *Indian J. Ent.*, 56(3), 280-291.

Singh, U. C. and S. V. Dhamdhere, 1989. Field screening of some guava varities against the bark eating caterpillar, *Indarbela quadrinotala* Walker. *Indian J. Ent.*, 51(2), 216-217.

Singh, A.V. 1984. Ecology and management of citrus leat miner, *Phillocnistis citrella* Stainton. M.Sc. Thesis, G.B. Pant University of Agriculture and Technology, Pantnagar, 106.

Singh, H. K. and H. N. Singh, 1990. Acetylcholinesterase and carboxylesterase sensitivity to organophosphate poisoning during different developmental stages of *Dysdercus koenigii* (F.). *Indian J. Ent.,* 52 (1), 50-56.

Singh, H. N. and Byas, 1973. Host selection in the tobacco caterpillar *Spodoptera litura* (Fab.) (Lepidoptera : Noctuidae). *Indian J. Agri. Sci.,* 43 (4), 357-360.

Singh, J. P. and D. Singh, 1991. Emergence pattern of Pink bollworm moth from overwintering larvae. *Indian. J. Ent.,* 53 (3), 487-789.

Singh, M.P. 1984. Field screening of some jujuble cultivars against the attack of back eating caterpillar, *Indarbela quadrinotata* Walker. *Madras agric. J.,* 71 (6), 416-417.

Singh, P. 1977. Artificial diets for insects, mites and spiders. IFI Plenum, New York.

Singh, S. and S.S. Sehgal, 1993. Studies on consumption and utilization of different food plants by larvae of *Spilosoma obliqua* Walker. *Indian J. Ent.,* 55 (1), 72-82.

Smith, J. B. 1904. Remarks on the catalogue of the Noctuidae in the collection of British Museum. *J. New York. Ento. Soc.,* 12(2), 93-104.

Smith, J. B. 1891. Contributions towards a monograph of the insect of the Lepidopterous family Noctuidae of temperate, North America Revision of the genus *Agrotis. Bull. U. S. Nath. Mus.,* 88, 1-245.

Solomon, J. D. 1962. Characters for determing sex in elm spanworm pupae. *J. Econ. Ent.,* 55, 269-270.

Srivastava, A.S., Lat, J. and H. P. Saxena, 1963. Note on the nature of damage, occurrence and incidence of insect pests of groundnut in U.P. *Labdev J. Sci. and Tech.,* 3, 141-143.

Stanev, M. and A. Kaitazov, 1962. Studies on the bionomics and ecology of the potato tuber moth Gnorimochema *(Phthorimaea operculella* Zeller) in Bugaria and means for its control. Izv. *Naik, Inst. Zashi, Rast.,* 3, 49-89.

Stark, R.W. 1959. Population dynamics of the lodgepole needle miner, *Nicrophorus* spp and the mite *Poecilochirus necrophori. J. Animi. Ecol.,* 37, 417-474. (W.L. 25559).

Stebbing, E. P. 1903. *Pyrausta machaeralis* Wlk. Departmental notes. II, XVIII, XIX : 301 – 311. Dept. of Ent. Banglore.

Subrahmanyam, B., Rao, P. J. and L. D. Tiwari, 1980. Juvenomimetic effects of diflubenzuron on *Achaea janata* Linn. *Indian J. Exp. Biol.,* 18(3), 324-325.

Sugi, S. 1987. Redescription of *Micreremites japonica* Sugi, Noctudiae and description of its new ally from Okinawa. *Japan Heterocerists Journal,* 141, 241 – 243.

Suryawanshi, D.S., Pawar, V. M. and P. S. Borikar, 2001. Life fecundity tables of *Earias vittella* Fabricius on okra and cotton. *Indian J. Ent.,* 25(4), 249 – 262.

Swinhoe, C. 1895. A list of the Lepidoptera of the Khasi Hills. *Trans. R. Ent. Soc. London,* 1-75.

Swinhoe, C. 1891. New species of Heterocera from the Khasi Hills, Part I. *Trans. R. Ent. Soc., London,* 473-495.

Swinhoe, C. 1919a. On the geographical distribution of *Cosmophila* of nottuid of family Genopteridae. *Ann. Mag. Nat. Hist. London,* 3, 309-313, Pls.IX, X.

Swinhoe, C.1889. On the new Indian Lepidoptera, chiefly Heterocera, *Proc. Zool. Soc. London*, 396-436.

Swinhoe, C. 1919b. Indo-Malayan and Australian Noctuidae. *Ann. Nat Hist; London,* 4, *US-127.*

Swihoe, C. 1986. On the Lepidoptera collected at Kurrachee. *Proc. zool. Soc. London,* 525-529.

Tamaki, G., Turner, J.E. and R.L. Wallis, 1972. Life tables for evaluating the rearing of the Zebra caterpillar. *J. Econ. Ent.,* 65, 1024-2027.

Tara, J. S. 1983. Investigation on the insect pests of mulberry (*Morus* sp.) in Jammu region of J&K. Ph. D. thesis, Univ. of Jammu, Jammu.

Taylor, L. R. 1984. Assessing and interpreting the spatial distribution of insect population. *Ann. Rev. Ent.,* 29, 321-327.

Thakur, M. L. and S.R.M. Pillai, 1985. Insect fauna of Subabul, *Leucaena leucocephala* Lam. from South India. *Indian forester,* 111(2), 68-77.

Thobi, V. V. 1961. Growth potential of *Prodenia litura* F. in relation to certain food plants at Surat. *Indian J. Ent.,* 23(4), 262-264.

Thompson, W.R. 1924. La theorie mathematique deiaction des parasites entomophages et. le. facteur du hasar du hasar d. *Annls Fal. Sci. Morseilla,* 2, 69-89.

Thorsteinson, A. J. 1960. Host selection in phytophagous insects. *Ann. Rev. Ent.,* 5, 193-218.

Tiwari, L. D. 1985. Effect of Dimilin on consumption and utilization of dry matter and dietary constituents of castor, *Ricinus communis* Linn. by the castor semilooper, *Achaea janata* Linn. *Indian J. Ent.,* 47(1), 9-13.

Tiwari, S. N. and A. K. Bhattacharya, 1987. Formulation of artificial diets for Bihar hairy catterpiller, *Spilosoma obliqua* Walker (Lepidoptera : Arctiidae). *Mem. Ent. Soc. India* No. 12. Division of Entomology, I.A.R.I. New Delhi, India.

Tiwari, S.N., Rathore, Y.S., Bhattacharya, A. K. and G.C. Sachan, 1988. Susceptibiity of several varieties of groundnut to *Spilosoma obliqua* Walker (Lepidoptera : Arctiidae). *Indian J. Ent.,* 50(2), 179-184.

Todd, E.L. 1983. The noctuid type material of John B. Smith (Lepidoptera). United states. Department of Agriculture, *Technical Bulletin,* 1645, 228.

Toetia, T. P.S. and V.S. Singh, 1968. On the oviposition behaviour and development and *Sitophilus oryzae* Linn. in various natural foods. *Indian J. Ent.,* 30(2),119-124.

Turunen, S. 1977. Food utilization and esterase activity in *Pieris brassicae* during chronic exposure to lindane containing food. *Ent. exp. appl.,* 21, 254-260.

Tutt, J. W. 1896. On the structural affinities of the genus *Demas. Can. Ent.,* 28, 81-83.

Tutt, J. W. 1902. British moths, XIII : pp. 368, London.

Tutt, J. W. and J. B. Smith, 1895. Catalogue of the Lepidopterous superfamily Noctuidae found in Bercal. *Am. Ent. Rec. J. VarL, 6,* 69-72.

Tutt, J.W. 1891-1892. The British noctuidae and their varieties, 1, XVI + 164, pp. 2, xviii : pp. 180. 3, 295, 4, 94, London.

Vaishampayan, S.M.and A. Bahadur, 1983. Seasonal activity of adults of teak defoliator *Hyblaea puera* and teak skeletonizer *Pyrausta machaeralis* (Pyralidae: Lepidoptera) monitored light by trap catch. Insect interrelations in forest and agro ecosystems / edited by P.K. Sen-Sarma, S.K. Kulshrestha, S.K. Sangal. Publication: Dehra Dun, India : Jugal Kishore, Doc. Type: Book chapter

Vanderzant, E.S. 1974. Development, significance and application of artificial diets for insects. *Ann. Rev. Ent.,* 19, 139 -160.

Vartak, V. D., Suryanarayan and B. V. Shetty, 1981. Biodiversity in the western Ghats: 2-1,1-4.

Vasuki, V. 1990. Ovipositional responses of vector mosquitoes to the IGR. Treated water. *Entomon,* 16 (3), 175 -178.

Veen, K. H. 1968. Recherches suv la maladie, due and metarhizium anisopliae chez le criquest pelerine. Mededelingen Landbouwhogeschool, Wagengingen (Thesis).

Vishwa Nath, Nath, T.N. and P J. Rao, 1983. Effect of sumithion on consumption and utiliza-tion of food by *Schistocerca gregaria* Forsk. *Indian J. Ent.,* 42, 174-179.

Wagner, D.L. 2005. Caterpillers of Eastern North America : A guide to identification and natural history. Princeton University Press, Princeton, N.I., 512.

Waldbauer, G. P., 1964. The consumption, digestion and utilization of Solonaceous and non-Solananceous plants by the larvae of tobacco hornworm *Protopurce sexta* Johan (Lepidoptera : Sphingidae). *Ent. exp. appl.,* 7, 253-263.

Waloff, N. 1968. Studies on the insect fauna of Scotch broom *Sarothamnus scoparins* (L). Wimmer. *Advances in Ecological Research,* 5, 87-208.

Watson, A. and D.T. Goodger, 1986. Catalogue of the Neotropical tiger moths, Occasional Papers on Systematic Entomology, 1, 1-71.

Watson, T. F. 1964. Influence of host plant conditions on population increase of *Tetranychus telarius* (Linnaeus). *Hilgardia,* 35, 275-322.

Wesley, W. K. 1956. Major insect pests of vegetables in Allahabad U. P. and their control. *Fmr.,* 30, 121-128.

William, C. 2009. Tiger moths and woolly bears : Behavior, ecology, and evolution of the Arctiidae. Oxford University Press: New York. ISBN 978-0-19-532737-3

Yadav, S. R. and M.M. Bachulkar, 1995. New plant records from Satara district. Part I. *Botanical Reporter J.,* 14 (1 and 2), 1-6.

Yadav, S.R. and M.M. Sardesai, 2002. Flora of Kolhapur District. Pub. Shivaji Uni., Kolhapur, India. pp.1-680.

Yamamoto, R.T. and G. Fraenkel, 1960. The suitability of tobacco for growth of the cigarette beetle *Lasioderma serricorne. J. Econ. Ent.,* 56, 381-384.

Zecevic, D. 1958. Daily food consumption of gypsy moth caterpillars on Dak trees and *Pyracantha coccinea* (in Serbo-Creation). *Zast. Bilja,* 50, 23-24.

Zoological Survey of India, 1983. Threatened Animals of India, Calcutta : Zoological Survey of India, pp. 307.

Index